Chicago Public Library

REFERENCE

Form 178 rev. 1-94

U·X·L

COMPLETE
LIFE
SCIENCE
RESOURCE

U·X·L
COMPLETE
LIFE
SCIENCE
RESOURCE

volume TWO: F-N

LEONARD C. BRUNO
JULIE CARNAGIE, EDITOR

AN IMPRINT OF THE GALE GROUP

DETROIT · SAN FRANCISCO · LONDON
BOSTON · WOODBRIDGE, CT

U·X·L Complete Life Science Resource

LEONARD C. BRUNO

Staff

Julie L. Carnagie, *U·X·L Senior Editor*

Carol DeKane Nagel, *U·X·L Managing Editor*

Meggin Condino, *Senior Market Analyst*

Margaret Chamberlain, *Permissions Specialist*

Randy Bassett, *Image Database Supervisor*

Robert Duncan, *Imaging Specialist*

Pamela A. Reed, *Image Coordinator*

Robyn V. Young, *Senior Image Editor*

Michelle DiMercurio, *Art Director*

Evi Seoud, *Assistant Manager, Composition Purchasing and Electronic Prepress*

Mary Beth Trimper, *Manager, Composition and Electronic Prepress*

Rita Wimberley, *Senior Buyer*

Dorothy Maki, *Manufacturing Manager*

GGS Information Services, Inc., *Typesetting*

Bruno, Leonard C.
  U·X·L complete life science resource / Leonard C. Bruno; Julie L. Carnagie, editor.
  p. cm.
  Includes bibliographical references.
  Contents: v. 1. A-E   v. 2. F-N   v. 3. O-Z.
  ISBN 0-7876-4851-5 (set)   ISBN 0-7876-4852-3 (vol. 1)   ISBN 0-7876-4854-X (vol. 2)
  1. Life sciences Juvenile literature. [1. Life sciences Encyclopedias.] I. Carnagie, Julie. II. Title.
  QH309.2.B78 2001   00-56376

U·X·L, an Imprint of the Gale Group
27500 Drake Rd.
Farmington Hills, MI 48331-3535

Printed in the United States of America

10 9 8 7 6 5 4 3 2 1

Table of Contents

Contents

volume THREE: O–Z

Reader's Guide

U·X·L Complete Life Science Resource explores the fascinating world of the life sciences by providing readers with comprehensive and easy-to-use information. The three-volume set features 240 alphabetically arranged entries, which explain the theories, concepts, discoveries, and developments frequently studied by today's students, including: cells and simple organisms, diversity and adaptation, human body systems and life cycles, the human genome, plants, animals, and classification, populations and ecosystems, and reproduction and heredity.

The three-volume set includes a timeline of scientific discoveries, a "Further Information" section, and research and activity section. It also contains 180 black-and-white illustrations that help to bring the text to life, sidebars containing short biographies of scientists, a "Words to Know" section, and a cumulative index providing easy access to the subjects, theories, and people discussed throughout *U·X·L Complete Life Science Resource.*

Acknowledgments

Special thanks are due for the invaluable comments and suggestions provided by the *U·X·L Complete Life Science Resource* advisors:

- Don Curry, Science Teacher, Silverado High School, Las Vegas, Nevada

- Barbara Ibach, Librarian, Northville High School, Northville, Michigan

- Joel Jones, Branch Manager, Kansas City Public Library, Kansas City, Missouri

- Nina Levine, Media Specialist, Blue Mountain Middle School, Peekskill, New York

Comments and Suggestions

We welcome your comments on this work as well as your suggestions for topics to be featured in future editions of *U·X·L Complete Life Science Resource*. Please write: Editors, *U·X·L Complete Life Science Resource*, U·X·L, 27500 Drake Rd., Farmington Hills, MI 48331-3535; call toll-free: 1-800-877-4253; fax: 248-699-8097; or send e-mail via www.galegroup.com.

Introduction

U·X·L Complete Life Science Resource is organized and written in a manner to emphasize clarity and usefulness. Produced with grades seven through twelve in mind, it therefore reflects topics that are currently found in most textbooks on the life sciences. Most of these alphabetically arranged topics could be described as important concepts and theories in the life sciences. Other topics are more specific, but still important, subcategories or segments of a larger concept.

Life science is another, perhaps broader, term for biology. Both simply mean the scientific study of life. All of the essays included in *U·X·L Complete Life Science Resource* can be considered as variations on the simple theme that because something is alive it is very different from something that is not. In some way all of these essays explore and describe the many different aspects of what are considered to be the major characteristics or signs of life. Living things use energy and are organized in a certain way; they react, respond, grow, and develop; they change and adapt; they reproduce and they die. Despite this impressive list, the phenomenon that is called life is so complex, awe-inspiring, and even incomprehensible that our knowledge of it is really only just beginning.

This work is an attempt to provide students with simple explanations of what are obviously very complex ideas. The essays are intended to provide basic, introductory information. The chosen topics broadly cover all aspects of the life sciences. The biographical sidebars touch upon most of the major achievers and contributors in the life sciences and all relate in some way to a particular essay. Finally, the citations listed in the "For Further Information" section include not only materials that were used by the author as sources, but other books that the ambitious and curious student of the life sciences might wish to consult.

Timeline of Significant Developments in the Life Sciences

c. 50,000 B.C.   *Homo sapiens sapiens* emerges as a conscious observer of nature.

c. 10,000 B.C.   Humans begin the transition from hunting and gathering to settled agriculture, beginning the Neolithic Revolution.

c. 1800 B.C.   Process of fermentation is first understood and controlled by the Egyptians.

c. 350 B.C.   Greek philosopher Aristotle (384–322 B.C.) first attempts to classify animals, considers nature of reproduction and inheritance, and basically founds the science of biology.

A.D. 1543   Flemish anatomist Andreas Vesalius (1514–1564) publishes *Seven Books on the Construction of the Human Body* which corrects many misconceptions regarding the human body and founds modern anatomy.

1615   The modern study of animal metabolism is founded by Italian physician, Santorio Santorio (1561–1636), who publishes *De Statica Medicina* in which he is the first to apply measurement and physics to the study of processes within the human body.

12,000 B.C.	3,000 B.C.	552
The dog is domesticated from the wolf	The world's population reaches 100,000	Buddhism reaches Japan

15,000 B.C.	7,500 B.C.	A.D. 1	1,000

1628	The first accurate description of human blood circulation is offered by English physician William Harvey (1578–1657), who also founds modern physiology.
1665	English physicist Robert Hooke (1635–1703) coins the word "cell" and develops the first drawing of a cell after observing a sliver of cork under a microscope.
1669	Entomology, or the study of insects, is founded by Dutch naturalist Jan Swammerdam (1637–1680), who begins the first major study of insect microanatomy and classification.
1677	Dutch biologist and microscopist Anton van Leeuwenhoek (1632–1723) is the first to observe and describe spermatozoa (sperm). He later goes on to describe different types of bacteria and protozoa.
1727	English botanist Stephen Hales (1677–1761) studies plant nutrition and measures water absorbed by roots and released by leaves. He states that the plants convert something in the air into food, and that light is a necessary part of this process, which later becomes known as photosynthesis.
1735	Considered the father of modern taxonomy, Swedish botanist Carl Linnaeus (1707–1778) creates the first scientific system for classifying animals and plants. His system of binomial nomenclature establishes generic and specific names.
1779	Dutch physician Jan Ingenhousz (1739–1799) shows that carbon dioxide is taken in and oxygen is given off by plants during photosynthesis. He also states that sunlight is necessary for this process.
1802	The word "biology" is coined by French naturalist Jean-Baptiste Lamarck (1744–1829) to describe the new science of living things. He later proposes the first scientific, but flawed, theory of evolution.

1650
England's first coffee house opens

1710
The first copyright law is established in Britain

1770
The Boston Massacre occurs

1620	1680	1740	1800

1809   Modern invertebrate zoology is founded by French naturalist Jean-Baptiste Lamarck (1744–1829) who also introduces the term "invertebrate."

1827   A mammalian egg is discovered by Estonian biologist Karl Ernst von Baer (1792–1876). He states that the human egg is not fundamentally different from that of other animals.

1831   English naturalist Charles Robert Darwin (1809–1882), begins his historic voyage on the H.M.S. *Beagle* (1831–36).

1839   German physiologist Theodore Schwann (1810–1882) states that all living things are made up of cells, each of which contains certain essential components. Schwann's theory is applied to both animals and plants and becomes known as the cell theory.

1858   Modern biology begins as German pathologist Rudolph Virchow (1821–1902) founds cellular pathology with his historic statement that "Every cell comes from a cell."

1859   The landmark book, *On the Origin of Species,* is published by Charles Darwin. This revolutionary work proposes a theory of evolution based on variation and survival of the fittest.

1864   Pasteurization is invented by French chemist Louis Pasteur (1822–1895). Earlier he recognized the relation between microorganisms and disease as well as microorganisms and fermentation.

1866   The laws of inheritance, or genetics, are first stated by Austrian botanist Gregor Johann Mendel (1822–1884). He also states that both male and female contribute equal factors (genes) to the offspring and that these factors do not blend but remain distinct.

1820	1840	1860
The Spanish Inquisition ends	The brass saxophone is invented	The internal combustion engine is patented

1810   1830   1850   1870

1873	Italian histologist Camillo Golgi (1843–1926) devises a way to stain tissue samples with inorganic dye and applies this new method to nerve tissues.
1882	German bacteriologist Robert Koch (1843–1910) establishes the classic method of preserving, documenting, and studying bacteria.
1882	German anatomist Walther Flemming (1843–1905) becomes the first to observe and describe mitosis or splitting of chromosomes, the structure in the cell that carries the cell's genetic material.
1900	Different types of human blood are discovered by Austrian American physician Karl Landsteiner (1868–1943), who names them A, B, AB, and O.
1901	Spanish histologist Santiago Ramon y Cajal (1852–1911) demonstrates that the neuron is the basis of the nervous system.
1902	Hormones are first named and understood by English physiologists Ernest H. Starling (1866–1927) and William H. Bayliss (1860–1924), who describe them as chemicals that stimulate an organ from a distance.
1905	English biochemist Frederick Gowland Hopkins (1861–1947) provides proof that "essential amino acids" cannot be manufactured by the body and must be obtained from food.
1907	Russian physiologist Ivan Pavlov (1849–1936) conducts pioneering studies on inborn reflexes and the conditioning of animals.
1910	American geneticist Thomas Hunt Morgan (1866–1945) works with the fruit fly *Drosophila* and establishes the chromosome theory of inheritance. This theory states that chromosomes are composed of discrete entities called genes that are the actual carriers of specific traits.

1880	1900	1920
Thomas Edison receives patent for the light bulb	Sigmund Freud pioneers psychoanalysis	Suffrage for American women becomes effective

1875	1890	1905	1920

1912	English biochemist Frederick Gowland Hopkins (1861–1947) proves that "accessory substances," later called vitamins, are essential for health and growth.
1932	German biochemist Hans Krebs (1900–1981) discovers that glucose (sugar) is broken down in a chain of reactions that comes to be called the Krebs cycle.
1953	The double helix structure of deoxyribonucleic acid (DNA) is discovered by American biochemist James Dewey Watson (1928–) and English biochemist Francis Harry Compton Crick (1916–). Their model explains how DNA transmits hereditary traits in living organisms, and forms the basis for all genetic discoveries that follow. This is considered one of the greatest of all scientific discoveries.
1961	Messenger ribonucleic acid (mRNA), which transfers genetic information to the ribosomes where proteins are made, is discovered by French biologists Jacques Lucien Monod (1910–1976) and Francois Jacob (1920–).
1978	The first "test tube" baby is born in England. Physicians remove an egg from the mother's ovary, fertilize it with the father's sperm in a petri dish, and reimplant it in the mother's uterus.
1982	A gene from one mammal (a rat growth hormone gene) functions for the first time in another mammal (a mouse). As a result, the mouse grows to twice its normal size.
1983	American biologist Lynn Margulis (1938–) discovers that cells with nuclei can be formed by the synthesis of non-nucleated cells (those without a nucleus, like bacteria).
1987	Genetically engineered plants are first developed.

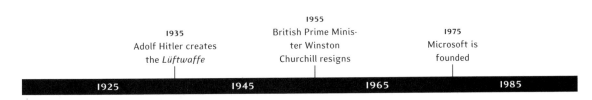

1935
Adolf Hitler creates
the *Lüftwaffe*

1955
British Prime Minister Winston
Churchill resigns

1975
Microsoft is
founded

1925   1945   1965   1985

1990	The Human Genome Project is established in Washington, D.C., as an international team of scientists announces a plan to compile a "map" of human genes.
1991	The gender of a mouse is changed at the embryo stage.
1992	The United Nations Conference on Environment and Development is held in Brazil and is attended by delegates from 178 countries, most of whom agree to combat global warming and to preserve biodiversity.
1995	The first complete sequencing of an organism's genetic make up is achieved by the Institute for Genomic Research in the United States. The institute uses an unconventional technique to sequence all 1,800,000 base pairs that make up the chromosome of a certain bacterium.
1997	The first successful cloning of an adult mammal is achieved by Scottish embryologist Ian Wilmut (1944–), who clones a lamb named Dolly from a cell taken from the mammary gland of a sheep.
1998	The first completed genome of an animal, a roundworm, is achieved by a British and American team. The genetic map shows the 97,000,000 genetic letters in correct sequence, taken from the worm's 19,900 genes.
1999	Danish researchers find what they believe is evidence of the oldest life on Earth—fossilized plankton from 3,700,000,000 years ago.
2000	Gene therapy succeeds unequivocally for the first time as doctors in France add working genes to three infants who could not develop their own complete immune systems.

1992		1995		1999
Bill Clinton becomes president of the United States		The Million Man March takes place		The first nonstop around-the-world balloon trip is made

| 1990 | 1993 | 1996 | 1999 |

Words to Know

A

Abiotic: The nonliving part of the environment.

Absorption: The process by which dissolved substances pass through a cell's membrane.

Acid: A solution that produces a burning sensation on the skin and has a sour taste.

Acid rain: Rain that has been made strongly acidic by pollutants in the atmosphere.

Acquired characteristics: Traits that are developed by an organism during its lifetime; they cannot be inherited by offspring.

Active transport: In cells, the transfer of a substance across a membrane from a region of low concentration to an area of high concentration; requires the use of energy.

Adaptation: Any change that makes a species or an individual better suited to its environment or way of life.

Adrenalin: A hormone released by the body as a result of fear, anger, or intense emotion that prepares the body for action.

Aerobic respiration: A process that requires oxygen in which food is broken down to release energy.

AIDS: A disease caused by a virus that disables the immune system.

Algae: A group of plantlike organisms that make their own food and live wherever there is water, light, and a supply of minerals.

Allele: An alternate version of the same gene.

Alternation of generations: The life cycle of a plant in which asexual stages alternate with sexual stages.

Amino acids: The building blocks of proteins.

Amoeba: A single-celled organism that has no fixed shape.

Amphibians: A group of vertebrates that spend part of their life on land and part in water; includes frogs, toads, and salamanders.

Anaerobic respiration: A stage in the breaking down of food to release energy that takes place in the absence of oxygen.

Anaphase: The stage during mitosis when chromatids separate and move to the cell poles.

Angiosperms: Flowering plants that produce seeds inside of their fruit.

Anther: The male part of a flower that contains pollen; a saclike container at the tip of the stamen.

Antibiotics: A naturally occurring chemical that kills or inhibits the growth of bacteria.

Antibody: A protein made by the body that locks on, or marks, a particular type of antigen so that it can be destroyed by other cells.

Antigen: Any foreign substance in the body that stimulates the immune system to action.

Arachnid: An invertebrate that has four pairs of jointed walking legs.

Arthropod: An invertebrate that has jointed legs and a segmented body.

Atom: The smallest particle of an element.

Autotroph: An organism, such as a green plant, that can make its own food from inorganic materials.

Auxins: A group of plant hormones that control the plant's growth and development.

Axon: A long, threadlike part of a neuron that conducts nerve impulses away from the cell.

B

Bacteria: A group of one-celled organisms so small they can only be seen with a microscope.

Binomial nomenclature: The system in which organisms are identified by a two-part Latin name; the first name is capitalized and identifies the genus; the second name identifies the species of that genus.

Biological community: A collection of all the different living things found in the same geographic area.

Biological diversity: A broad term that includes all forms of life and the ecological systems in which they live.

Biomass: The total amount of living matter in a given area.

Biome: A large geographical area characterized by distinct climate and soil and particular kinds of plants and animals.

Biosphere: All parts of Earth, extending both below and above its surface, in which organisms can survive.

Biotechnology: The alteration of cells or biological molecules for a specific purpose.

Bipedalism: Walking on two feet; a human characteristic.

Binary fission: A type of asexual reproduction that occurs by splitting into two more or less equal parts; bacteria usually reproduce by splitting in two.

Blood: A complex liquid that circulates throughout an animal's body and keeps the body's cells alive.

Blood type: A certain class or group of blood that has particular properties.

Brain: The control center of an organism's nervous system.

Breeding: The crossing of plants and animals to change the characteristics of an existing variety or to produce a new one.

Bud: A swelling or undeveloped shoot on a plant stem that is protected by scales.

C

Calorie: A unit of measure of the energy that can be obtained from a food; one calorie will raise the temperature of one kilogram of water by one degree Celsius.

Camouflage: Color or shape of an animal that allows it to blend in with its surroundings.

Carbohydrates: A group of naturally occurring compounds that are essential sources of energy for all living things.

Carbon cycle: The process in which carbon atoms are recycled over and over again on Earth.

Carbon dioxide: A major atmospheric gas.

Carbon monoxide: An odorless, tasteless, colorless, and poisonous gas.

Carnivores: A certain family of mammals that have specially shaped teeth and live by hunting.

Carpel: The female organ of a flower that contains its stigma, style, and ovary.

Cartilage: Smooth, flexible connective tissue found in the ear, the nose, and the joints.

Catalyst: A substance that increases the speed at which a chemical reaction occurs.

Cell: The building block of all living things

Cell theory: States that the cell is the basic building block of all life-forms and that all living things, whether plants or animals, consist of one or more cells.

Cellulose: A carbohydrate that plants use to form the walls of their cells.

Central nervous system: The brain and spinal cord of a vertebrate; it interprets messages and makes decisions involving action.

Centriole: A tiny structure found near the nucleus of most animal cells that plays an important role during cell division.

Cerebellum: The part of the brain that coordinates muscular coordination and balance; the second largest part of the human brain.

Cerebrum: The part of the brain that controls thinking, speech, memory, and voluntary actions; the largest part of the human brain.

Cetacean: A mammal that lives entirely in water and breathes air through lungs.

Chlorophyll: The green pigment or coloring matter in plant cells; it works by transferring the Sun's energy in photosynthesis.

Chloroplast: The energy-converting structures found in the cells of plants.

Chromatin: Ropelike fibers containing deoxyribonucleic acid (DNA) and proteins that are found in the cell nucleus and which contract into a chromosome just before cell division.

Chromosome: A coiled structure in the nucleus of a cell that carries the cell's deoxyribonucleic acid (DNA).

Cilia: Short, hairlike projections that can beat or wave back and forth; singular, cilium.

Classification: A method of organizing plants and animals into categories based on their appearance and the natural relationships between them.

Cleavage: Early cell division in an embryo; each cleavage approximately doubles the number of cells.

Cloning: A group of genetically identical cells descended from a single common ancestor.

Cnidarian: A simple invertebrate that lives in the water and has a digestive cavity with only one opening.

Cochlea: A coiled tube filled with fluid in the inner ear whose nerve endings transmit sound vibrations.

Community: All of the populations of different species living in a specific environment.

Conditioned reflex: A type of learned behavior in which the natural stimulus for a reflex act is substituted with a new stimulus.

Consumers: Animals that eat plants who are then eaten by other animals.

Cornea: The transparent front of the eyeball that is curved and partly focuses the light entering the eye.

Cranium: The dome-shaped, bony part of the skull that protects the brain; it consists of eight plates linked together by joints.

Crustacean: An invertebrate with several pairs of jointed legs and two pairs of antennae.

Cytoplasm: The contents of a cell, excluding its nucleus.

D

Daughter cells: The two new, identical cells that form after mitosis when a cell divides.

Decomposer: An organism, like bacteria and fungi, that feed upon dead organic matter and return inorganic materials back to the environment to be used again.

Dendrite: Any branching extension of a neuron that receives incoming signals.

Deoxyribonucleic acid (DNA): The genetic material that carries the code for all living things.

Differentiation: The specialized changes that occur in a cell as an embryo starts to develop.

Diffusion: The movement or spreading out of a substance from an area of high concentration to the area of lowest concentration.

Dominant trait: An inherited trait that masks or hides a recessive trait.

Double helix: The "spiral staircase" shape or structure of the deoxyribonucleic acid (DNA) molecule.

E

Ecosystem: A living community and its nonliving environment.

Ectoderm: In a developing embryo, the outermost layer of cells that eventually become part of the nerves and skin.

Ectotherm: A cold-blooded animal, like a fish or reptile, whose temperature changes with its surroundings.

Element: A pure substance that contains only one type of atom.

Endangered species: Any species of plant or animal that is threatened with extinction.

Endoderm: In a developing embryo, the innermost layer of cells that eventually become the organs and linings of the digestive, respiratory, and urinary systems.

Endoplasmic reticulum: A network of membranes or tubes in a cell through which materials move.

Endotherm: A warm-blooded animal, like a mammal or bird, whose metabolism keeps its body at a constant temperature.

Energy: The ability to do work.

Enzyme: A protein that acts as a catalyst and speeds up chemical reactions in living things.

Epidermis: The outer layer of an animal's skin; also the outer layer of cells on a leaf.

Eukaryote: An organism whose cells contain a well-defined nucleus that is bound by a membrane.

Eutrophication: A natural process that occurs in an aging lake or pond as it gradually builds up its concentration of plant nutrients.

Evolution: A scientific theory stating that species undergo genetic change over time and that all living things originated from simple organisms.

Exoskeleton: A tough exterior or outside skeleton that surrounds an animal's body.

Extinction: The dying out and permanent disappearance of a species.

F

Fermentation: A chemical process that breaks down carbohydrates and other organic materials and produces energy without using oxygen.

Fertilization: The union of male and female sex cells.

Fetus: A developing embryo in the human uterus that is at least two months old.

Flagella: Hairlike projections possessed by some cells that whip from side to side and help the cell move about; singular, flagellum.

Food chain: A sequence of relationships in which the flow of energy passes.

Food web: A network of relationships in which the flow of energy branches out in many directions.

Fossil: The preserved remains of a once-living organism.

Fruit: The mature or ripened ovary that contains a flower's seeds.

Fungi: A group of many-celled organisms that live by absorbing food and are neither plant nor animal.

G

Gaia hypothesis: The idea that Earth is a living organism and can regulate its own environment.

Gamete: Sex cells used in reproduction; the ovum or egg cell is the female gamete and the sperm cell is the male gamete.

Gastric juice: The digestive juice produced by the stomach; it contains weak hydrochloric acid and pepsin (which breaks down proteins).

Gene: The basic unit of heredity.

Genetic code: The information that tells a cell how to interpret the chemical information stored inside deoxyribonucleic acid (DNA).

Genetic disorder: Conditions that have some origin in a person's genetic makeup.

Genetic engineering: The deliberate alteration of a living thing's genetic material to change its characteristics.

Genetic theory: The idea that genes are the basic units in which characteristics are passed from one generation to the next.

Genetic therapy: The process of manipulating genetic material either to treat a disease or to change a physical characteristic.

Genotype: The genetic makeup of a cell or an individual organism; the sum total of all its genes.

Geolotic record: The history of Earth as recorded in the rocks that make up its crust.

Germination: The earliest stages of growth when a seed begins to transform itself into a living plant that has roots, stems, and leaves.

Gland: A group of cells that produce and secrete enzymes, hormones, and other chemicals in the body.

Golgi body: A collection of membranes inside a cell that packages and transports substances made by the cell.

Greenhouse effect: The name given to the trapping of heat in the lower atmosphere and the warming of Earth's surface that results.

Gymnosperm: Plants with seeds that are not protected by any type of covering.

H

Habitat: The distinct, local environment where a particular species lives.

Heart: A muscular pump that transports blood throughout the body.

Hemoglobin: A complex protein molecule in the red blood cells of vertebrates that carries oxygen molecules in the bloodstream.

Herbivore: Animals that eat only plants.

Herpetology: The scientific study of amphibians and reptiles.

Heterotroph: An organism, like an animal, that cannot make its own food and must obtain its nutrients be eating plants or other animals.

Hibernation: A special type of deep sleep that enables an animal to survive the extreme winter cold.

Homeostasis: The maintenance of stable internal conditions in a living thing.

Hominid: A family of primates that includes today's humans and their extinct direct ancestors.

Hormones: Chemical messengers found in both animals and plants.

Host: The organism on or in which a parasite lives.

Hybrid: The offspring of two different species of plant or animal.

Hypothesis: A possible answer to a scientific problem; it must be tested and proved by observation and experiment.

I

Ichthyology: The branch of zoology that deals with fish.

Immunization: A method of helping the body's natural immune system be able to resist a particular disease.

Inbreeding: The mating of organisms that are closely related or which share a common ancestry.

Instincts: A specific inborn behavior pattern that is inherited by all animal species.

Interphase: The stage during mitosis when cell division is complete.

Invertebrates: Any animal that lacks a backbone, such as paramecia, insects, and sea urchins.

Iris: The colored ring surrounding the pupil of the vertebrate eye; its muscles control the size of the pupil (and therefore the amount of light that enters).

K

Karyotype: A diagnostic tool used by physicians to examine the shape, number, and structure of a person's chromosomes when there is a reason to suspect that a chromosomal abnormality may exist.

L

Lactic acid: An organic compound found in the blood and muscles of animals during extreme exercise.

Larva: The name of the stage between hatching and adulthood in the life cycle of some invertebrates.

Lipids: A group of organic compounds that include fats, oils, and waxes.

Lysosome: Small, round bodies containing digestive enzymes that break down large food molecules into smaller ones.

M

Malnutrition: The physical state of overall poor health that can result from a lack of enough food to eat or from eating the wrong foods.

Mammals: A warm-blooded vertebrate with some hair that feeds milk to its young.

Medulla: The part of the brain just above the spinal cord that controls certain involuntary functions like breathing, heartbeat rate, sneezing, and vomiting; the smallest part of the brain.

Meiosis: A specialized form of cell division that takes place only in the reproductive cells.

Membrane: A thin barrier that separates a cell from its surroundings.

Mendelian laws of inheritance: A theory that states that characteristics are not inherited in a random way but instead follow predictable, mathematical patterns.

Mesoderm: In a developing embryo, the middle layer of cells that eventually become bone, muscle, blood, and reproductive organs.

Metabolism: All of the chemical processes that take place in an organism when it obtains and uses energy.

Metamorphosis: The extreme changes that some organisms go through when they pass from an egg to an adult.

Metaphase: The stage during mitosis when the chromosomes line up across the center of the spindle.

Microorganism: Any form of life too small to be seen without a microscope, such as bacteria, protozoans, and many algae; also called microbe.

Migration: The seasonal movement of an animal to a place that offers more favorable living conditions.

Mineral: An inorganic compound that living things need in small amounts, like potassium, sodium, and calcium.

Mitochondria: Specialized structures inside a cell that break down food and release energy.

Mitosis: The division of a cell nucleus to produce two identical cells.

Molars: Chewing teeth that grind or crush food; the back teeth in the jaws of mammals.

Molecule: A chemical unit consisting of two or more linked atoms.

Mollusk: A soft-bodied invertebrate that is often protected by a hard shell.

Molting: The shedding and discarding of the exoskeleton; some insects molt during metamorphosis, and snakes shed their outer skin in order to grow larger.

Monerans: A group of one-celled organisms that do not have a nucleus.

Mutation: A change in a gene that results in a new inherited trait.

N

Natural selection: The process of survival and reproduction of organisms that are best suited to their environment.

Neuron: An individual nerve cell; the basic unit of the nervous system.

Niche: The particular job or function that a living thing plays in the particular place it lives.

Nitrogen cycle: The stages in which the important gas nitrogen is converted and circulated from the nonliving world to the living world and back again.

Nucleic acid: A group of organic compounds that carry genetic information.

Nutrients: Substances a living thing needs to consume that are used for growth and energy; for humans they include fats, sugars, starches, proteins, minerals, and vitamins.

Nutrition: The process by which an organism obtains and uses raw materials from its environment in order to stay alive.

O

Omnivore: An animal that eats both plants and other animals.

Organ: A structural part of a plant or animal that carries out a certain function and is made up of two or more types of tissue.

Organelle: A tiny structure inside a cell that performs a particular function.

Organic compound: Substances that contain carbon.

Organism: Any complete, individual living thing.

Ornithology: The branch of zoology that deals with birds.

Osmosis: The movement of water from one solution to another through a membrane or barrier that separates the solutions.

Oviparous: Term describing an animal that lays or spawns eggs which then develop and hatch outside of the mother's body.

Ovoviparous: Term describing an animal whose young develop inside the mother's body, but who receive nourishment from a yolk and not from the mother.

Oxidation: An energy-releasing chemical reaction that occurs when a substance is combined with oxygen.

Ozone: A form of oxygen found naturally in the stratosphere or upper atmosphere that shields Earth from the Sun's harmful ultraviolet radiation.

P

Paleontology: The scientific study of the animals, plants, and other organisms that lived in prehistoric times.

Parasite: An organism that lives in or on another organism and benefits from the relationship.

Ph: A number used to measure the degree of acidity of a solution.

Phenotype: The outward appearance of an organism; the visible expression of its genotype.

Pheromones: Chemicals released by an animal that have some sort of effect on another animal.

Photosynthesis: The process by which plants use light energy to make food from simple chemicals.

Physiology: The study of how an organism and its body parts work or function normally.

Pistil: The female part of a flower made up of organs called carpels; located in the center of the flower, parts of it become fruit after fertilization.

Plankton: Tiny, free-floating organisms in a body of water.

Pollen: Dustlike grains produced by a flower's anthers that contain the male sex cells.

Pollution: The contamination of the natural environment by harmful substances that are produced by human activity.

Population: All the members of the same species that live together in a particular place.

Predator: An organism that lives by catching, killing, and eating another organism.

Primate: A type of mammal with flexible fingers and toes, forward-pointing eyes, and a well-developed brain.

Producer: A living thing, like a green plant, that makes its own food and forms the beginning of a food chain, since it is eaten by other species.

Prokaryote: An organism, like bacteria or blue-green algae, whose cells lack both a nucleus and any other membrane-bound organelles.

Prophase: The stage in mitosis when the chromosomes condense or, coil up, and the sister chromatids become visible.

Protein: The building blocks of all forms of life.

Protozoa: A group of single-celled organisms that live by taking in food.

Pseudopod: A temporary outgrowth or extension of the cytoplasm of an amoeba that allows it to slowly move.

R

Radioactive dating: A method of determining the approximate age of an old object by measuring the amount of a known radioactive element it contains.

Recessive trait: An inherited trait that may be present in an organism without showing itself. It is only expressed or seen when partnered by an identical recessive trait.

Reptiles: A cold-blooded vertebrate (animal with a backbone) with dry, scaly skin and which lays sealed eggs.

Respiration: A series of chemical reactions in which food is broken down to release energy.

Retina: The lining at the back of the eyeball that contains nerve endings or rods sensitive to light.

Rh factor: A certain blood type marker that each human blood type either has (Rh-positive) or does not have (Rh-negative).

Rhizome: A creeping underground plant stem that comes up through the soil and grows new stems.

Ribonucleic acid (RNA): An organic substance in living cells that plays an essential role in the construction of proteins and therefore in the transfer of genetic information.

Rods: Nerve endings or receptor cells in the retina of the eye that are sensitive to dim light but cannot identify colors.

S

Sap: A liquid inside a plant that is made up mainly of water and which transports dissolved substances throughout the plant.

Sedimentation: The settling of solid particles at the bottom of a body of water that are eventually squashed together by pressure to form rock.

Smooth muscle: Muscle that appears smooth under a microscope; they are involuntary muscles since they cannot be controlled.

Sponge: An invertebrate that lives underwater and survives by taking in water through a system of pores.

Spontaneous generation: The incorrect theory that nonliving material can give rise to living organisms.

Spore: Usually a single-celled structure with a tough coat that allows an organism, like bacteria or fungi, to reproduce asexually under the proper conditions.

Stamen: The male organ of a flower consisting of a filament and an anther in which the pollen grains are produced.

Stigma: The tip of a flower's pistil upon which pollen collects during pollination and fertilization.

Stimulus: Anything that causes a receptor or sensory nerve to react and carry a message.

Stomata: The pores in leaves that allow gases to enter and leave; singular, stoma.

Stress: A physical, psychological, or environmental disturbance of the well-being of an organism.

Striated muscle: Muscle that appears striped under a microscope; also called skeletal muscles, they are under the voluntary control of the brain.

Symbiosis: A relationship between two different species who benefit by living closely together.

Synapse: The space or gap between two neurons across which a nerve impulse or a signal is transmitted.

T

Taxonomy: The science of classifying living things.

Telophase: The near-final phase of mitosis in which the cytoplasm of the dividing cell separates two sets of chromosomes.

Territory: An area that an animal claims as its own and which it will defend against rivals.

Tissue: The name for a group of similar cells that have a common structure and function and which work together.

Toxins: Chemical substances that destroy life or impair the function of living tissue and organs.

Transpiration: Loss of water by evaporation through the stomata of the leaves of a plant.

Tropism: The growth of a plant in a certain manner or direction as a response to a particular stimulus, such as when a plant grows toward the light source.

V

Vacuole: A bubble-like space or cavity inside a cell that serves as a storage area.

Variation: The natural differences that occur between the individuals in any group of plants or animals; if inherited, these differences are the raw materials for evolution.

Vascular plants: Plants with specialized tissue that act as a pipeline for carrying the food and water they need.

Vegetative reproduction: The asexual production of new plants from roots, underground runners, stems, or leaves.

Vertebrates: Animals that have a backbone and a skull that surrounds a well-developed brain.

Virus: A package of chemicals that infects living cells.

Vitamins: Organic compounds found in food that all animals need in small amounts.

Viviparous: Term describing an animal whose embryos develop inside the body of the female and who receive their nourishment from her.

Z

Zygote: A fertilized egg cell; the product of fertilization formed by the union of an egg and sperm.

Research and Activity Ideas

Activity 1: Studying an Ecosystem

Ecosystems are everywhere—your backyard, a nearby park, or even a single, rotting log. To study an ecosystem, you need only choose an individual natural community to observe and study and then begin to keep track of all of the interactions that occur among the living and nonliving parts of the ecosystem. Look carefully and study the entire ecosystem, deciding on what its natural boundaries are. Making a map or a drawing on graph paper of the complete site always helps. Next, you should classify the major biotic (living) and abiotic (nonliving) factors in the ecosystem and begin to observe the organisms that live there. Binoculars sometimes help to observe distant objects or to keep from interfering with the activity. A small magnifying glass is also useful for studying small creatures. You should also search for evidence of creatures that you do not see. A camera is also useful sometimes, especially when comparing the seasonal changes in an ecosystem. It is very important to keep a notebook of your observations, keeping track of any creatures you find and where you find them. You can learn more about your ecosystem by counting the different populations discovered there, as well as classifying them according to their ecosystem roles like producer, consumer, or decomposer. A diagram can then be made of the ecosystem's food web. You can search for evidence of competition as well as other types of relationships such as predator-prey or parasitism. You can even keep a record of changes such as plant or animal growth, the birth of offspring, or weather fluctuations. Finally, you can try to predict what might happen if some part of the ecosystem were disturbed or greatly changed. Ecosystems themselves are related to other ecosystems in many ways, and it is important to always realize that all the living and nonliving things on Earth are ultimately connected to one another.

Activity 2: Studying the Greenhouse Effect

The greenhouse effect is the name given to the natural trapping of heat in the lower atmosphere and the warming of Earth's surface that results. This global warming is a natural process that keeps our planet warm and hospitable to life. However, when this normal process is exaggerated or enhanced because of certain human activities, too much heat can be trapped and the increased warming could result in harmful climate changes.

The greenhouse effect can be produced by trying the following experiment. Using two trays filled with moist soil and some easy-to-grow seeds like beans, place a flat thermometer on the soil surface of each tray. After inserting tall wooden skewers in the four corners of one tray, cover it completely with plastic wrap and secure it with a large rubber band. Leave the other tray uncovered and place both trays outside where they are sheltered from the rain but exposed to the Sun. Record the temperature of each tray at the same time each day and note all the differences between the plants. The plastic-wrapped tray should be warmer and its seedling plants should grow larger. This is evidence of the beneficial aspects of the greenhouse effect. However, if the plastic wrap is left over the seedlings for too long they will overheat, wither, and die.

Activity 3: Studying Photosynthesis

If you have ever picked up a piece of wood that has been sitting on the grass for some time and noticed that the patch underneath has lost its greenness and appears yellow or whitish, you have witnessed the opposite of photosynthesis. Since a green plant cannot exist without sunlight, when it is left totally in the dark, the chlorophyll departs from its leaves and photosynthesis no longer takes place. The key role of sunlight can be easily demonstrated by germinating pea seeds and placing them in pots of soil. After placing some pots in a place where they will receive plenty of direct sunlight, place the other pots in a very dark area. After a week to ten days, compare the seedlings in the sunlight to those left in the dark. The root structure of both is especially interesting.

Another way of demonstrating the importance of sunlight to a plant is to pick a shrub, tree, or houseplant that has large individual leaves. Using aluminum foil or pieces of cardboard cut into distinct geometrical shapes that are small enough not to cover the entire leaf but large enough to cover at least half, paperclip each shape to a different leaf. After about a week, remove the shapes from the leaves and compare what you see now to those leaves that were not covered. The importance of sunlight will be dramatically noticeable.

Finally, as a way of demonstrating the exchange of gases (carbon dioxide and oxygen) that occurs during photosynthesis, place a large glass over some potted pea seedlings and place them in sunlight. In time, you will notice that some liquid has condensed on the inside of the glass. This condensation is water vapor that has been given off by the plant when it exchanges oxygen for the carbon dioxide it needs.

Activity 4: Studying Osmosis

In the life sciences, osmosis occurs at the cellular level. For example, in mammals it plays a key role in the kidneys, which filter urine from the blood. Plants also get the water they need through osmosis that occurs in their root hairs. Everyday examples of osmosis can be seen when we sprinkle sugar on a grapefruit cut in half. We notice that the surface becomes moist very quickly and a sweet syrup eventually forms on its top surface. Once the crystallized sugar is dissolved by the grapefruit juices and becomes a liquid, the water molecules will automatically move from where they are greater in number to where they are fewer, so the greater liquid in the grapefruit forms a syrup with the dissolved sugar. Placing a limp stalk of celery in water will restore much of its crispness and gives us another example of osmosis.

Osmosis occurs in plants and animals at the cellular level because their cell membranes are semipermeable (meaning that they will allow only molecules of a certain size or smaller to pass through them). Osmosis can be studied directly by observing how liquid moves through the membrane of an egg. This requires that you get at an egg's membrane by submerging a raw egg (still in its shell) completely inside a wide-mouth jar of vinegar. Record the egg's weight and size (length and diameter) before doing this. The acetic acid in the vinegar will eventually dissolve the shell because the shell is made of calcium carbonate or limestone which reacts with acid to produce carbon dioxide gas. You will observe this gas forming as bubbles on the surface. After about 72 hours, the shell should be dissolved but the egg will remain intact because of its transparent membrane.

After carefully removing the egg from the jar of vinegar, weigh and measure the egg again. You will notice that its proportions have increased. The egg has gotten larger because the water in the vinegar moved through the egg's membrane into the egg itself (because of the higher concentration of water in the vinegar than in the egg). The contents of the egg did not pass out of the membrane since the contents is too large.

The opposite of this activity can be performed using thick corn syrup instead of water. If the egg has its shell removed in the same manner as above but is then immersed for about 72 hours in a jar of syrup, you will find that the egg will have shrunken noticeably. This is because the water concentration of the syrup outside the egg is much less than that inside the egg, so the membrane allows water to move from the egg to the syrup.

Activity 5: Studying Inherited Traits

An inherited trait is a feature or characteristic of an organism that has been passed on to it in its genes. This transmission of the parents' traits to their offspring always follows certain principles or laws. The study of how these inherited traits are passed on is called genetics. Genetics influences everything about us, including the way we look, act, and feel, and some of our inherited traits are very noticeable. Besides these very obvious traits like hair and skin color, there are certain other traits that are less noticeable but very interesting. One of these is foot size. Another is free or attached earlobes. Still another is called "finger hair."

All of these are traits that are passed from parents to their offspring. You can collect data on any particular inherited characteristic and therefore learn more about how genetics works. You will need to collect data about each trait and develop a chart. Any of the above inherited traits can be analyzed. For example, there are generally two types of earlobes. They may be free, and therefore hang down below where the earlobe bottom joins the head, or they may be attached and have no curved bottom that appears to hang down freely. Foot length is simply the size of your own foot and is measured from the tip of the big toe to the back of the heel. The finger hair trait always appears in one of two forms. It is either there or it isn't. People who have the finger hair trait have some hair on the middle section of one or more fingers (which is the finger section between the two bendable joints of your finger).

In order to study one of these interesting traits like finger hair or type of earlobes, you should construct a table or chart that records data on the trait for as many of your family members as you wish. Although it is best to include a large sampling, such as starting with both sets of grandparents and working through any aunts, uncles, and cousins you can contact, even a small sample with only a few members can be helpful. Once you have determined the type of trait each family member has, you should draw your family's "pedigree" for that trait. This is simply a diagram of connected individuals that looks like any other genealogical diagram (which starts at the top with two parents and draws a line from them down to their offspring, and so on). You should use some sort of easily identifiable code or color to signify which individual has or does not have a certain trait. The standard coding technique for tracing the occurrence of a trait in a family is to represent males by squares and females by circles. Usually, a solid circle or square means that a person has the trait, while an empty square or circle shows they do not. In more elaborate pedigrees, a half-colored circle or square means that the person is a carrier but does not show the trait. Once you have done your pedigree, you may do the same for a friend's family and compare his or her family's distribution of the same trait. By comparing the two families' pedigrees for the same trait, you may be able to find certain general patterns of inheritance and to answer certain basic questions. For example, in studying the finger hair trait, you may be able to answer the question whether or not both parents must have finger hair for their offspring to also have it. You might also discover whether both parents having finger hair means that *every* offspring must show the same trait.

U·X·L

COMPLETE
LIFE
SCIENCE
RESOURCE

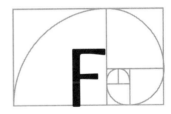

Family

The term family is one of the seven major classification groups that biologists use to identify and categorize living things. These seven groups are hierarchical or range in order of size. Family is located between the groups order and genus. The classification scheme for all living things is: kingdom, phylum, class, order, family, genus, and species.

Since the grouping family includes organisms that are even more alike than those in the group order, it is not always obvious which should or should not be included. One example that is easy to distinguish would be that of cat and dog. Both belong to the kingdom Animalia, the phylum Chordata, the class Mammalia, and the order Carnivora, but each is placed in a different family. Since the dog has nonretractable claws and hunts its prey by chasing it, it belongs to the family Canidae. The cat has retractable claws and hunts by stalking and surprise, and therefore belongs to the family Felidae.

For plants, all family names end in -aceae, while for animals, the family names end in -idae. The grouping family is certainly important to biologists, but in most cases, they often find that stating a genus and species is more than sufficient for consistent understanding and identification of an organism under consideration.

[*See also* **Class; Classification; Genus; Kingdom; Order; Phylum; Species**]

Fermentation

Fermentation is a chemical process that breaks down carbohydrates and other organic materials and produces energy without using oxygen. This process is carried out by microorganisms such as bacteria, molds, and fungi. Alcohol fermentation is a well-known type of fermentation where sugar is broken down into alcohol and carbon dioxide.

Fermented products have been used by people for thousands of years, primarily to make the alcohol in beer and wine and to make bread dough

LOUIS PASTEUR

One of the most extraordinary scientists in history, French chemist and microbiologist (a person specializing in the study of microorganisms) Louis Pasteur (1822–1895) is considered the founder of microbiology. He also contributed to our understanding of fermentation (a chemical process that breaks down carbohydrates and other organic materials and produces energy without using oxygen), developed the germ theory of disease, improved immunization, and proved that heating kills microorganisms (an organism of microscopic size). This process of using heat was named pasteurization after the famed scientist.

Louis Pasteur was born in Dole, France, and his family moved to Arbois when he was very young. He attended school there and appeared to be a mediocre student in every way. Still, he stayed in school despite near poverty, and after attending a lecture in chemistry and being inspired by it, he decided to study this new and fascinating subject. Studying chemistry, he suddenly became an excellent student, and by the age of twenty-six earned his Ph.D. and made a major discovery concerning crystals for which he won a national award.

By 1854, the "mediocre" student had become dean of the Faculty of Sciences at the University of Lille, and was asked by the French wine industry to help them with their spoilage problem. Very often, wine and beer spoiled, or went sour, as they aged, ruining tons of good beverage without anyone knowing why or what to do. Pasteur took up the problem and discovered almost immediately with his microscope that the yeast (various single-celled fungi capable of fermenting carbohydrates) in sour wine had an elongated shape, while the yeast in good wine was spherical, or round. When the good (round) yeast ferments, it produces alcohol. When the bad (elongated) yeast ferments, it produces lactic acid (a syrupy liquid). He suggested heating the wine or beer gently at about 120°F (48.9°C) after it had been properly made. Pasteur stated that the heat would kill any yeast left,

rise. Although they did not understand what made it happen, the ancient Egyptians knew that if they allowed bread dough to stand for several hours, it became lighter and better tasting than if baked immediately. What they did not know was that the dough was lightened by the carbon dioxide gas produced by the fermentation of sugar. This happened not because the Egyptians knew enough to add yeast (a single-celled fungus) to the dough, but because leaving the dough uncovered allowed microscopic organisms like yeast and bacteria to float in on the breeze and break down the dough's sugars into alcohol and carbon dioxide. The carbon dioxide gas then became trapped in the dough and made it rise, while the alcohol

especially the bad yeast, and if the wine were properly corked, it would not go sour. Heating wine seemed barbaric to the French, but they tried the experiment and it worked. Ever since, a gentle heating that kills unwanted microorganisms has been called pasteurization. Besides beer and wine, milk also is now pasteurized.

Saving the French wine industry made Pasteur a hero, so it is not surprising that its silk industry asked him to do the same thing for them. He did by showing them how to get rid of a killer parasite (an often harmful organism that lives on or in a different organism) that was killing the silk worms. This led Pasteur to work with communicable (contagious) diseases. He had long felt that disease was something that was caused by unseeable organisms and then was spread person-to-person. By now, Pasteur had considerable experience using his microscope to identify different kinds of microorganisms such as bacteria and fungi. So when he decided to work on what is now called the germ theory of disease, he was following one of his favorite sayings, "Chance favors the prepared mind." Pasteur then developed techniques for culturing (growing) and examining several disease-causing bacteria. He identified both *Staphylococcus* and *Streptococcus,* which cause serious, sometimes fatal infections, and also cultured the bacteria that cause cholera. It was in working with these infectious bacteria that Pasteur realized that weakening them allows them to be used as a vaccine. From this discovery, he developed a vaccine for the disease anthrax, as well as one for rabies, a deadly disease contracted from the bite of an infected, rabid animal. If any individual had achieved one or two of these accomplishments, he or she would be considered among the pioneers in the life sciences. Yet Pasteur did all this and more. His germ theory of disease is considered by many scientists as the single most important medical discovery of all time because it not only showed doctors how to fight and prevent disease, but it supplied the all-important correct theory that would guide future research. Pasteur is truly one of the giants of biology.

would evaporate during baking. The Egyptians also discovered that by allowing certain grains like barley to begin to spoil, they could obtain a drink with a pleasing side effect (alcohol). The same effect could be achieved by allowing grapes to spoil since grapes contain yeast that grow naturally on their skins.

Throughout history, the process of fermentation was shrouded in mystery and superstition. During the seventeenth century, the English chemist Robert Boyle (1627–1691) correctly predicted that an understanding of the fermentation process would lead to the discovery of the causes of other phenomena like disease. Boyle's prediction came true when the French chemist, Louis Pasteur (1822–1895), proved that yeast caused fermentation in beer and wine. After this discovery, Pasteur turned his research toward the spread of diseases caused by other microorganisms.

Pasteur's work saved France's wine industry, which could not understand why its burgundy wine was spoiling. Pasteur discovered that wine normally contained yeast cells that produced alcohol. However, he also realized that wine containing bacteria and other microorganisms produced lactic acid when they fermented, and thus spoiled the wine. Pasteur showed that fermentation caused by living organisms is too small to be seen without a microscope, and that the end product of the fermentation process depends on both what is being fermented and what microbes are the catalyst (something that starts a chemical reaction). Pasteur taught France's wine industry how to kill unwanted bacteria by the gentle heating of the wine at about 120°F (48.9°C). This process is called pasteurization after the great scientist.

Today, fermentation is well understood and can be controlled. Fermentation is a large part of today's food industry, with some form of fermentation taking place in the production of many food products like yogurt, buttermilk, cheese, soy sauce, cured meats, pickled vegetables, and chocolate, as well as in alcoholic beverages and bread. In some cases, antibiotics and other medications can be produced by fermentation, as can ethyl alcohol that is added to gasoline to produce gasohol. Fermentation is also critical to today's disposal of solid waste by converting it to carbon dioxide, water, and mineral salts.

[*See also* **Bacteria; Carbohydrates; Carbon Dioxide; Fungi**]

Fertilization

Fertilization is the union of male and female sex cells. Also called conception in humans, fertilization occurs during sexual reproduction that necessarily involves two parents. If fertilization is successful, a pregnancy occurs that results in the creation of a unique, new individual.

Fertilization is the key moment of sexual reproduction. It is the moment that begins the combining of the genes (hereditary characteristics) of two separate individuals to produce a unique, new offspring. For fertilization to occur, the male sperm and the female egg, or ovum, must be brought together or come together physically under the proper conditions. One they meet and actually touch, the process of fertilization can begin. This process has two stages. In the first stage, the ovum is "activated" by the contacting sperm, setting in motion a series of chemical reactions. The second stage is the actual fusion or uniting of the male gamete (sex cell) with the female gamete.

FERTILIZATION FOR AQUATIC ANIMALS

Before either stage can happen in an animal's reproductive system, other important things must take place. Sexual reproduction in animals requires that male sperm swim to the egg. For animals that already live in a watery environment (like fish and frogs), this is no problem, and they usually engage in external fertilization. In this process, the female lays her eggs or releases them in the water while the male simultaneously deposits his sperm in the same area. Fertilization is left to chance as the female eggs release chemicals that attract the sperm.

FERTILIZATION FOR LAND ANIMALS

Among animals that live on land however, fertilization is usually done internally, meaning that the male deposits his sperm inside the body of the female. The egg is large and does not move, while the sperm are small and very mobile. Since sperm must swim to the egg, the male produces a fluid that transports them, while the female also provides additional internal fluid. Most male animals use a specialized organ called a penis to deposit their sperm safely inside the female.

Once the sperm reaches the egg and comes into physical contact with it, a chemical reaction begins. First the head of the individual sperm bursts open and releases a chemical that bores a hole into the outer covering of the egg. As this is happening, tiny membranes called "microvilli" emerge from both the sperm head and the egg, and actually join together or fuse. At this point, fertilization occurs and the sperm is engulfed in egg's cytoplasm (the jelly-like fluid in a cell). As the sperm continues deeper into the egg, the nuclei (the control center of a cell) of both eventually meet and form a new nucleus. Basically, the genetic contents of the sperm are joined to those of the egg. The now-fertilized egg is called a zygote. It is from this zygote that the entire, complete new individual will develop.

This process begins with division: first into two cells, then four, then eight, and so on. As soon as the division begins, the zygote becomes an embryo. It is from this single fertilized cell that a multicellular individual with tissues, organs, and organ systems comes to be. When an embryo has formed organs, it is then called a fetus. Also, as soon as penetration by one sperm occurs, the egg takes action to prevent penetration by any other sperm. It changes its electrical charge from negative to positive and creates a hard protective layer that cannot be penetrated.

ARTIFICIAL FERTILIZATION

For a long time fertilization was not understood. Now it is not only understood but it can be achieved artificially. Fertilization by artificial insemination is achieved in animals by introducing sperm-containing se-

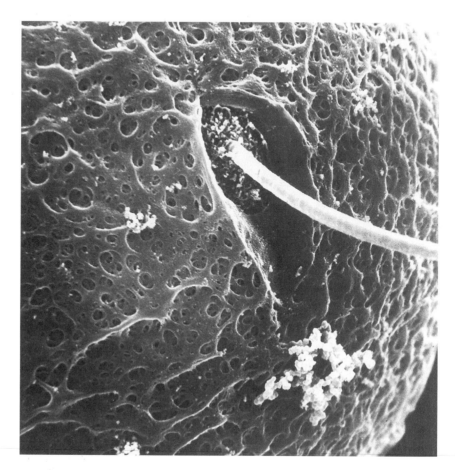

Sperm penetrating a hamster ovum. When the genetic material of the sperm and egg fuses, fertilization is complete. (Reproduced by permission of Photo Researchers, Inc. Photograph by David Phillips.)

men into the vagina or uterus of the female without any sexual contact. This process was first developed for breeding cattle and horses, and is sometimes used in humans as long as the male partner is able to produce sperm. Another form of artificial insemination is called "in vitro" fertilization. This involves mixing the sperm and ovum in a laboratory dish and transplanting the fertilized egg into the woman's uterus where it will develop normally. Children born through this form of artificial insemination were often called "test-tube babies." The first such test-tube baby was Mary Louise Brown, born in England in 1978. The procedure was used successfully in the United States for the first time in 1981, and since then more than 50,000 American babies have been born as a result of this technique.

[*See also* **Egg; Human Reproduction; Reproduction, Sexual; Sperm**]

Fish

A fish is a vertebrate (an animal with a backbone) that lives in the water and breathes through gills. It has a streamlined body and is usually protected by a hard coat of scales. Fish species vary greatly in size and shape, but all are cold-blooded and most have fins. Fish live in watery habitats as diverse as stagnant ponds and subzero polar water.

As a vertebrate that lives its entire life underwater, a fish is ectothermic or cold-blooded. This means that it is not necessarily cold but that its own, internal temperature rises or lowers to meet that of its environment. If anything defines a fish as a fish, it is that it breathes through gills. These respiratory organs that lie behind and to the side of the mouth are able to absorb oxygen that is dissolved in water and to give off carbon dioxide as the water passes over the gills' filaments. Oxygen-rich blood is then pumped by a heart to the rest of its body.

Most fish have a streamlined body over which water flows easily as it moves through the water. Fish are able to swim forward by contracting the muscles on each side of their body in turn so that their tail whips from side to side and pushes them forward. Fins allow them to maneuver and have control and balance, while an inflatable swim bladder keeps them from sinking when they are not swimming. As a fish descends to deeper waters, the increase in pressure compresses and deflates the bladder, allowing the fish to swim deeper. As the fish rises again and the pressure decreases, the bladder begins to inflate with gas. The majority of fish have scales that are overlapping plates that protect its body. A fish does not shed these scales, because they grow as its body grows. Mucus usually covers these scales, as it helps the fish to glide more easily through the water.

Most fish reproduce sexually through the union of male sperm and female eggs, but it takes place outside the female's body by what is called spawning. Spawning is the release of eggs by the female into the water. The male then releases his sperm over the eggs and some of the eggs are fertilized.

Biologists have grouped fish into three classes: jawless fish, cartilage fish, and bony fish. A jawless fish is a primitive, wormlike fish without a hinged jaw. This means that it usually has a simple, sucker-like mouth instead. A lamprey eel is a good example of this ancient type of fish. The lamprey is a parasite and sucks the blood and juices from live fish. The only other jawless fish is the scavenger fish called the hagfish. It has a round mouth and attaches itself to the bodies of dead or dying fish, feeding on the contents of the victim's body.

A close-up photo of a long nose gar showing its overlapping scales used to protect its body. (Reproduced by permission of Field Mark Publications. Photograph by Robert J. Huffman.)

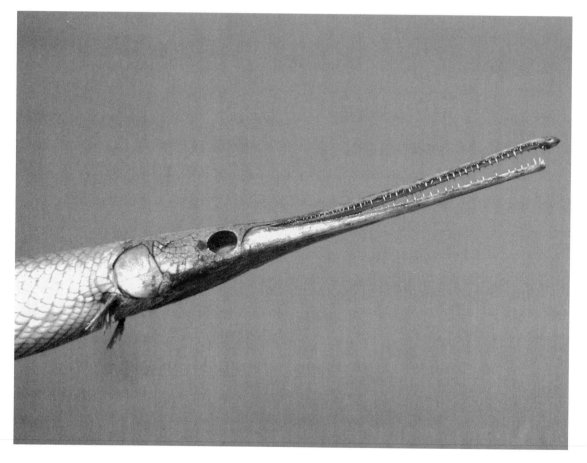

Cartilage fish have an endoskeleton (internal skeleton) made entirely of strong, flexible cartilage instead of bone. These fish are almost always hunters such as sharks, skates, and rays who live in the ocean. They all have jaws, scales, and paired fins. Their skin is covered with tiny scales that feel like sandpaper. The shark is an especially ferocious predator that must swim all the time because it has no swimbladder. Sharks are powerful swimmers and have teeth that grow in rows that move forward to replace lost teeth.

Bony fish make up the third and largest category of fish. Also called "true fish," these have an endoskeleton made up mostly of bone. The familiar trout, salmon, cod, and sardine are all bony fish. Bony fish all have a gill cover or flap and like all fish, a keen sense of smell. Bony fish usually have highly maneuverable fins that allow them to make rapid and complex movements. Although bony fish are, as their name implies, all bone, their skeleton is in fact very light and thin because they use the natural buoyancy of water to support their bodies. Fish are an important food source for people.

[*See also* **Ichthyology**]

Flower

A flower is a plant structure that contains the organs needed for sexual reproduction. The function of a flower is to make seeds. Flowers and the seeds they produce are also a major source of nutrients for almost all animals.

THE STRUCTURE OF FLOWERS

Most plants have flowers, and any plant that produces some sort of flower, even a small, colorless one, is a flowering plant. Grasses and oak trees are flowering plants just like roses and cherry trees. As many as 200,000 types of flowers have been classified, from tiny pondweed flowers to the bathtub-size flower of the tropical Giant Rafflesia. Despite the enormous variety in size, shape, color, and fragrance, all flowers have similar structure. Every flower has four basic organs or parts: the sepals, the petals, the stamen, and the pistil.

The Sepals and Petals. The sepals are the outermost part of the flower (where the flower emerges from the stem), and resemble green flaps that protect the flower when it is still a bud. The petals are usually the flower's most distinctive part and signal to animal pollinators (animals who transfer pollen, containing male sex cells to the pistil containing female sex cells) with bright colors and strong scents. Sometimes a flower's petals resemble a female insect, attracting males who pollinate as they land on their "mate." However, flowers that use the wind as a pollinator rather than animals usually have small petals or none at all.

The Stamen. The stamen is the male reproductive organ of a flower and lies inside the petals. Each is a slender stalk or a ribbon-like thread called a filament with an enlarged tip or head called an anther. The anther is like a small sac and contains the pollen, which are dustlike particles that contain the plant's male sex cells.

The Pistil. The pistil is the female reproductive organ of a flower and is usually located in the center of the flower. It is here that the seed is actually produced. Each pistil has three parts: the stigma, the style, and the

A close-up photograph of an open bleeding heart flower showing its petals, stamen, and pistil. (Reproduced by permission of Field Mark Publications. Photograph by Robert J. Huffman.)

ovary. The stigma is the sticky top of the style, which is a slender stalk or tube that connects with the ovary below. The ovary contains the ovules that store the female sex cells. In terms of functions, the stigma catches or collects the pollen, the style is the tube down which the pollen travels to the ovary, and the ovary is where the pollen fertilizes the ovules. It is the ovule that develops into the seed after fertilization, and the ovary that becomes the plant's fruit.

There are several other names for a flower's different subparts. For example, the protective petals around the stamens are collectively called the corolla. The sepals are also collectively called the calyx. In flowers that have two or more pistils (called compound pistils), the individual pistil is called a carpel. Flowers that have all these parts are termed perfect or complete, while those missing one or more parts are called incomplete or imperfect flowers.

REPRODUCTION IS THE KEY TO CLASSIFICATION

The flowering plants of the world have been classified into roughly three hundred families according to these flower parts. Thus, even though certain plants may grow in very different climates and soils and have varying shapes and colors, they may still be part of the same family because of how their reproductive organs look and function. Since reproduction is the key in classifying flowers, the lily family includes not only the tulip and hyacinth but also the onion, garlic, aloe, and yucca plant. All are pollinated by insects, contain three sepals and petals that closely resemble each other, six stamens, one pistil, long sheathlike leaves with parallel veins, and fruit that contains many seeds within one capsule. Further, the pea family includes not only beans, lentils, and peanuts but trees like the locust, vines like the wisteria, and herbs like licorice.

Flowers are critically important as the key to a flowering plant's ability to reproduce more of its own kind. Exactly when a flower blooms is also very important, since plants need to flower and form fruits and seeds before the cold season resumes. The length of daylight and darkness, as well as consistent temperature change, are mechanisms that signal a plant it should begin to blossom.

Besides acting as a reproductive agent for plants, flowers also beautify the world for people. Since ancient times, humans have prized flowers for their shapes, colors, and fragrances. They also have used flowers as medicines and given them symbolic meanings.

[*See also* **Botany; Plant Anatomy; Plant Reproduction**]

Food Chains and Webs

Food chains and food webs show how energy is transferred in an ecosystem (an area in which living things interact with each other and their environment) from one organism to another in the form of food. Since all organisms in an ecosystem need nutrients in the form of food, the terms food chain and food web describe the feeding relationship between an ecosystem's different populations.

Among the many processes that occur in every ecosystem, none is more important than the transfer of energy from one living thing to another. Without this energy exchange, no organisms would be able to survive. Ecologists use the term food chain to describe the typical path or route that energy (food) takes as it moves from one group of living things to another. A simple food chain would be: green plant to mouse to snake to eagle. In this example, the green plant is the first link in the chain, producing chemical energy from sunlight through a process of photosynthesis. The plant is eaten by the mouse, which absorbs the plant's energy. In turn, the mouse is eaten by the snake, who absorbs the mouse's energy.

An ecological or energy pyramid. The lowest level consists of producers, the next higher level of first-order consumers, the next higher level of second-order consumers, and so on. Note that the total number organisms found in any one level decreases as one goes up the pyramid. (Illustration courtesy of Gale Research.)

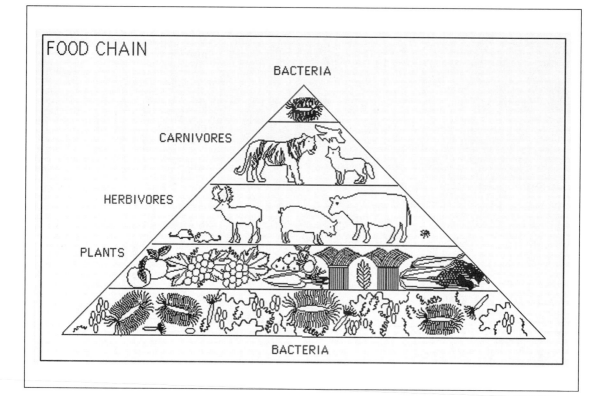

Finally, the eagle consumes the snake and obtains the snake's energy. This chain becomes circular when the eagle dies and its organic matter is reduced to nutrients, which are returned to the soil by organisms called decomposers.

Ecologists now prefer the term "food web" as a more realistic description of what really happens in this process. Since they feel that the word "chain" implies an orderly linking of equal parts, they use instead "food web" which can be described as a collection of food chains. In the natural world, food chains are extraordinarily complex, since there is no exact order stating which creature has to eat which. The notion of a web is more complicated than that of a chain and suggests a network of connections rather than a direct, one-to-one linking.

The first link in a food chain (or first stage of a food web) are the organisms known as primary producers. These are also called "autotrophs" since they can make their own food. Green plants and some forms of bacteria are primary producers since they begin the chain by performing photosynthesis. By capturing sunlight and using its energy to carry out a series of chemical reactions, green plants make glucose which is packed with energy. In the sea, primary producers are floating microorganisms known as plankton. On land, green plants are eaten by herbivores (plant-eating animals) who are considered to be the primary consumers. In turn, these primary consumers become food for other animals called secondary consumers. Each step up the ladder consists of fewer flesh-eating animals. Every food chain then begins with an autotroph and ends with a carnivore (flesh-eating animal) that itself is not eaten by a larger animal. The bear is a good example of a carnivore that is at the top of its food chain. When top carnivores die, their bodies are usually eaten by scavengers, like vultures. What remains is broken down by decomposers (bacteria and fungi) and returned to Earth as nutrients to be used by autotrophs (green plants).

Ecologists also use a model called an "energy pyramid" to depict the actual energy-transfer relationships in a food web. They have found that each species occupies a certain place or level on that pyramid that they call its "trophic level." Trophic refers to nourishment or nutrition, and a certain trophic level is the stage which a given organism occupies on the pyramid. In an energy pyramid, energy flows from one trophic level to the next, but only in one direction (bottom to top). At the base or bottom of the pyramid are the producers or plants, the most plentiful resource. Primary consumers are at the next level and are just below the secondary consumers. The pyramid is not just another way of showing a pattern, but demonstrates that the same amount of energy is never transferred up the pyramid. Studies have shown that the amount of energy passed on to the next higher pyramid level is only about 10 percent of the energy that the organism at the lower level received. This is because a great deal of energy is lost into the environment as heat, which cannot be reused. As energy passes up to the higher levels of the pyramid, less and less of it is left to reach the top. As a result, the total number of organisms that can participate in a food web is severely limited. It also shows how important the producers at the bottom of the pyramid are, since the less energy they produce the less energy is passed on up the pyramid.

Ecologists use models like energy pyramids to evaluate how much energy an ecosystem can produce, and therefore how much life it can sustain. They usually try to measure the total productivity of the producers and turn that into an amount of heat energy per unit of area.

Another way of measuring productivity is to measure the "biomass" (total amount of organic matter produced) of an ecosystem. Certain ecosystems are naturally more productive than others. For example, a forest is more productive than a desert. Also, human activity can have a disruptive effect on a food web if it causes a major change at one of its levels.

[*See also* **Ecosystem**]

Forests

A temperate, or deciduous, forest is a geographic area usually dominated by deciduous trees (trees that lose their leaves before winter). These forests experience four separate seasons and are home to a wide variety of insects, birds, and animals. Only a small fraction of the original forests that once covered the world's temperate areas remains today.

Dense deciduous forests once covered most of the world's temperate zones. Found on the eastern side of North America, most of Europe (except for its mountains), eastern Asia, and parts of southwestern South America, they are named after their most characteristic feature, which is large trees that lose their leaves before winter. Today, in whatever natural forest remains, we can only get a hint of what this original deciduous forest must have looked like.

Some deciduous forests are referred to as "climax" forests. A climax community is a balanced, stable habitat that has reached its highest or most developed form. It is usually dominated by one type of plant species. For example, the few giant sequoia trees remaining today on the West Coast of North America once lived, and were dominant in, a climax community.

FORESTS ARE IMPORTANT TO THE ECOSYSTEM

Now, as in the past, forests not only support a wide range of other plant and animal life, but they play a vital role in many ecosystems (areas in which living things interact with each other and the environment). Forests maintain the structure and fertility of forest soil and prevent soil erosion and floods. Forests also absorb carbon dioxide and release oxygen through photosynthesis (the process by which plants use light energy to make their food). In this manner, forests help control the level of greenhouse gases, which, in turn, helps prevent global warming. Forest trees also provide a huge number of products upon which human society depends. We still build most of our homes out of wood and make most of our paper from wood pulp. Trees also still provide fuel and fibers that are used not only in construction but in making foods and medicines.

PLANT LIFE IN FORESTS

The dominant plant life in a temperate forest is the broad-leaved tree, so-called because its green leaves have a flat, broad shape. These deciduous trees drop their colorful leaves as winter begins and go dormant for the colder months. Trees cannot draw water from the soil if the temperature of the air is less than about 40°F (5°C). The leaves from deciduous trees fall to the forest floor and begin the process of recycling. Worms, insects, fungi, and other microscopic organisms slowly convert this fallen debris into humus, which builds up the minerals in the soil.

Like a tropical rain forest, a temperate forest also has several horizontal layers of activity (like floors in an apartment house). The tallest trees form a canopy, or umbrella, although nowhere near as dense as that of a rain forest. However, it still shades the next layer of shorter trees, which are often oddly shaped as they reach in any direction they can to get more sunlight. Both types of trees are home to climbing animals and birds. Below these shorter trees are woody shrubs and bushes, many of which have berries. These provide cover and food for birds as well as large mammals like bears that thrive on late summer berries. Under the woody bushes and shrubs are grasses, ferns, flowers, and herbs. A great amount of animal activity occurs at this level. At the very bottom is the forest floor where everything dead usually falls and is slowly decomposed, to be recycled and reused in another form.

ANIMAL LIFE IN THE FORESTS

Today, many of the larger carnivores (animals that eat other animals) have left the remaining temperate forests and gone north to the coniferous (evergreen) forests. These evergreen forests are also called boreal forests, or the taiga, and have long, cold, dry winters and short summers. However, black bears, deer, raccoons, and squirrels are still commonly seen in temperate forests. The richest or most productive of all the North American forests are those in the southern Appalachians where maple, beech, hickory, horse chestnut, and birch trees shelter rhododendron, wild cherry, and magnolia shrubs.

[*See also* **Biome; Rain Forest; Taiga**]

Fossil

Fossils are the preserved remains of a once-living organism. They form in many different ways and they can provide us with information on the climate, geology, and geography of ancient Earth. Fossils also provide strong evidence for evolution (the process by which living things change over generations).

Fossils are usually thought of as the remains of once-living things that have been mineralized or turned into rock. While this describes a fossil, the notion of fossilization includes several other methods of preserving the evidence of ancient life. The use of fossils to study ancient life and its development, or evolution, is called paleontology. The word fossil comes from the Latin word *fossilis,* meaning "something dug up." This is actually how most fossils are discovered. Most are found below the surface of Earth in a preserved form, since they had been covered up at one time or another. A fossil can be the partial or complete bodily remains of a plant or animal, or it can be the more common "trace fossil" which is more like some evidence of the organism's life or activities. Trace fossils are things like trails, coprolites (fossilized animal dung), or footprints.

THE FORMATION OF FOSSILS

Fossils can be formed in many ways, but they all have one process in common. They all replace the relatively fragile organic structure of a living thing with something that is harder and which lasts longer. This can only happen when certain conditions apply. Under normal conditions, if an animal falls dead on the ground, it will eventually begin to rot, or decompose and will be reduced to its basic organic compounds, which are then recycled. In the end, it completely and totally disappears. However, if the animal falls dead and slides into a tar pool, asphalt lake, or peat bog, it may be completely preserved because of the lack of oxygen or the absence of bacteria, which are both needed in order for decomposition to occur. In an extremely hot and dry climate, an animal may undergo mummification, or be rapidly dehydrated, under the right conditions; therefore, it will be preserved. In other extreme climates in the north, completely preserved specimens of woolly mammoths and even humans have been recovered after being frozen under glacial conditions for thousands of years. A final way that nature can capture and preserve a complete specimen is in amber (the sticky resin of a pine tree). Insects were usually fossilized this way when they became trapped inside the amber. When the resin itself became fossilized, the insect was forever frozen in a drop of what looks like yellow glass.

Although these are examples are more like preservation than true fossilization, they are all considered to be fossils. True fossils are the remains of once-living organisms that are preserved in stone. When geology (the study of the Earth) became more of a real science in the late eighteenth and early nineteenth centuries, geologists began exploring below Earth's surface. These scientists went deeper into successively older and older layers of what is called sedimentary rock (formed by the gradual settling of sediments). It was in these buried layers that they found entire communities of fossils in each layer. Often, the deeper layers of rock contained animals that were very different from any known animals, since the deeper the geologists dug, the older the fossil. These unknown animal bodies that were preserved were those that had been covered by sediments very quickly before they could start to decay. However, they needed to be more than just covered to become fossils. They also needed to be deprived of oxygen. This often happened when a dead animal sunk to the bottom of a body of water and was deeply covered by mud. As more sediments piled up and pressure in-

Fossil

The fossil remains of a baby dinosaur. This is considered a true fossil by scientists since it has been preserved in stone. (Reproduced by permission of AP/Wide World Photos.)

creased, the organic material that made up the animal was slowly replaced by minerals and the body eventually became a fossil made of stone. These type of fossils might be described as a replica of the real living thing.

In the rare process known as petrifaction, both the external shape and the internal structure of a plant or animal are perfectly reproduced. The Petrified Forest of Arizona contains trees that were buried by volcanic eruptions and underwent petrifaction. In this rare process, a molecule-by-molecule replacement occurred, with the end result being the replacement of natural wood fibers by silica. This replacement is usually so accurate that even the cell structure of the tree can be determined.

Paleontology, or the study of fossils, informs scientists about the organisms that lived on Earth long before human life evolved. It also tells us a great deal about Earth's climate in those ancient times. The oldest known fossils are 3,500,000-year-old bacteria. They represent not only the oldest known form of life on Earth but help scientists to learn more about the origins of life itself. Fossils also tell scientists about extinction that has taken place. Finally, most of what is known about evolution is based on the fossil record. It is fossils that provide the most reliable evidence for evolution since for some organisms, paleontologists are able to compare their fossils from different layers of rock (which are from different ages) and actually trace how they evolved physically.

[*See also* **Geologic Record**]

Fruit

Fruit refers to anything that contains seeds. From a botanical standpoint, a fruit is the mature or ripened ovary that contains a flower's seeds. Therefore, fruit may be dry and hard as well as soft and juicy, and many of the foods commonly thought of as vegetables are in fact fruit. Flowering plants produce fruit either to protect their seeds or to help the seed be better dispersed or scattered over a wide area.

As the seed-bearing part of a flowering plant, fruit develops and grows from a flower's ovaries. After a flower has been pollinated and its ovules (female sex cells) have been fertilized by a pollen grain (male sex cells), the flower begins to change. Since the flower no longer needs them, its stamens (male reproductive organs) and pistils (female reproductive organs) wither and the petals fall off. The ovary, a hollow structure located near the base of the flower that contains the ovules or female sex cells, starts to grow into fruit. The ovules become the seeds. The plant may produce a fruit containing one or more seeds, depending on the species.

Horticulturalists, or people who grow crops for commercial purposes, have a different way of defining fruit than do botanists (people specializing in the study of plants) since both plants and fruits contain seeds. To fruit growers, neither nuts nor cucumbers are considered to be a fruit, but they are definitely fruit to a botanist since they both contain seeds. Unlike the sometimes complicated rules used by horticulturalists, botanists classify a fruit by its structure, designating fruit as either simple or compound. A simple fruit is formed from a single ripened ovary. A compound fruit is the product of two or more ovaries. By far, the majority of fruit are simple, such as peaches, nuts, and berries. There are two types of simple fruit: fleshy or dry, depending upon their texture. Fleshy fruits are exactly what they sound like and come in three types: berries, drupes, and pomes. To botanists, berries include bananas and tomatoes since they have a completely fleshy ovary wall; drupes have a single pit or stone for a seed, like olives or peaches; and pomes have an inedible core, like an apple or a pear. Simple dry fruit include grains like corn and rice, as well as the more obvious pods of beans and peas. Compound fruits, which have developed from several ovaries, are fewer in number than simple fruit. There are two types of compound fruit: aggregate and multiple fruit. In an aggregate fruit, like an orange or a raspberry, the fruit developed from a single flower that had several ovaries. Multiple fruits are less common and have ovaries from several flowers. The fig and the pineapple are examples of multiple fruits that develop from a cluster of flowers on a single stem.

Fruit can also be described by the manner in which they disperse their seeds into the environment. Fleshy fruit ripen, fall to the ground, and decay, leaving their seed to possibly germinate (begin to grow or sprout) and start a new plant. Plants have evolved systems to avoid the overcrowding that would result if new plants grew only near the parent plant. Therefore many fleshy

Although many people think of tomatoes as a vegetable, they are really classified as a fruit since they contain seeds. (Reproduced by permission of Field Mark Publications. Photograph by Robert J. Huffman.)

fruits are edible and very tasty to animals that either spit out the seeds as they eat or consume the seeds with the fruit and eventually pass the undigested seeds in their feces, depositing them somewhere else. Nuts and other dry fruits are carried off by animals and buried for later eating. Those fruits left buried may germinate in the spring. Coconuts have a waterproof covering around their single seed, allowing them to be dispersed by the tides. Other dry fruits have burrs or thistles around their seeds that catch onto an animal's fur and are transported elsewhere for germination.

Fruit is very important to the human diet. Many of the fleshy fruits contain a high content of sugar and important vitamins needed by humans and other animals. Fruit also provides humans and other animals a means of obtaining the energy that plants have harnessed from the Sun. Without fruits, the human body lacks the nutrients and vitamins to fight off certain diseases.

Fungi

Fungi are a group of many-celled organisms that live by absorbing food and are neither plant nor animal. They are so different that biologists have given them their own separate kingdom among the five kingdoms or forms of life (monerans, protists, fungi, plants, and animals). Fungi play a key role as decomposers and recyclers, but they can also cause disease in plants and animals.

The kingdom Fungi is made up of yeasts, molds, and mushrooms. Most are many-celled organisms with a complex cell structure. Although some fungi resemble plants and have roots, they lack leaves and chlorophyll and cannot make their own food. Instead, they absorb their food directly from their surroundings, including living or dead matter. Fungi also digest their food outside their bodies. They do this by releasing enzymes onto their food, which breaks down the food for absorption. Fungi are different in the way they reproduce. Nearly all species of fungi reproduce asexually by forming special reproductive cells called spores. They do not produce embryos as plants and animals do. Instead, every fungus produces powdery spore cells that are so light they can be dispersed by the wind. These microscopic spores can resist harsh conditions and remain ready to germinate when conditions are right.

Fungi play an essential role in the cycles of nature because they break down organic matter like dead plants or animals and allow their basic nutrients to be recycled. Without fungi in the soil acting as nature's decomposers, we would be living in a sea of waste. When fungi eat, they break down or decompose organic matter into simple substances like carbon, nitrogen, and hydrogen, which become available for other living things to use. Fungi are usually too small to see, but when they form fruiting bodies (such as mushrooms growing on a rotting log), they become obvious. Those mushrooms we notice on a log are hard at work breaking down the dead wood.

Although beneficial, fungi can also be harmful to certain forms of life. This kind of fungi are parasitic and are known as biotrophs. Biotrophs are organisms that live by absorbing organic compounds from living matter. These fungi can destroy crops by attacking a plant's major systems, or they can contribute to plant diseases. Certain fungi, like molds, can cause foods to spoil, sneakers to smell, and toes to itch from athlete's foot. Other fungi, like yeast, are put to good use and are essential to making bread, wine, beer, and certain cheeses, since they cause the all-important fermentation process (the breaking down of carbohydrates into alcohol and carbon dioxide). In the twentieth century, scientists discovered that important drugs could be derived from fungi. Antibiotics—like penicillin as well as the wonder drug cyclosporin, which makes organ transplants possible—are good examples. Finally, mushrooms are cultivated like any crop and are sold commercially to be eaten as a food.

Magnified yeast cells. Yeast is an important fungi since it is essential in making products such as bread, cheese, beer, and wine. (Reproduced by permission of Photo Researchers, Inc.)

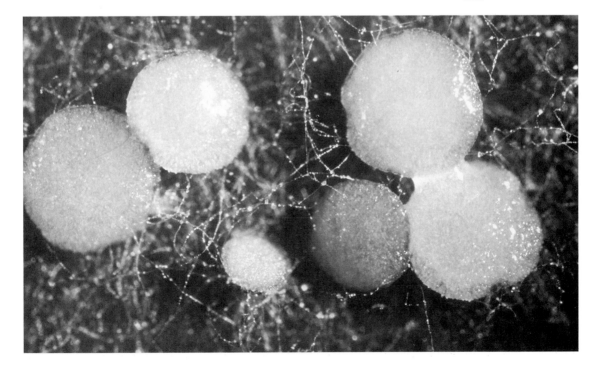

The main groups of fungi are chyrids, water molds, sporangium fungi, sac fungi, and club fungi. Chyrids live in muddy or aquatic habitats and feed on decaying plants, although some live as parasites. Water molds are important decomposers in watery environments. Sporangium fungi are commonly known as bread mold and are also found in soil and manure. Sac fungi, which comprise more than 30,000 species, include the yeast used to leaven bread and to make alcoholic beverages. Club fungi include the familiar mushroom as well as stinkhorns and puffballs. All fungi can reproduce asexually, and making spores is the most common method of reproduction. Spores are similar to seeds but much smaller and simpler since they usually contain only one or two cells. Fungi are able to respond to changes in their environment and can produce a large amount of spores in a short time if necessary (as in drought conditions). Most fungi release their spores into the air; the wind carries the ultralight spores into the atmosphere where they can travel great distances. If a spore lands on organic matter and conditions are good, it will germinate or sprout and produce a new fungus. With a typical mushroom, the part we notice above-ground is the spore-producing part of the organism. When a fungus is mature enough, a good rainfall will cause this spore-bearing part to swell and push aboveground, making mushrooms appear where there were none the day before.

A world with no fungi would be a world full of dead plants and animals that would neither rot and nor disappear naturally. It would be a world in constant need of basic substances since it could not break down organic matter and therefore could not recycle. It would also be a world without mushrooms. Mushroom farming is a large industry since people in all parts of the world eat mushrooms both raw and cooked. Wild mushrooms should never be eaten, however, since it takes an expert to know the difference between an edible mushroom and a poisonous one.

[*See also* **Antibiotics; Decomposition**]

Gaia Hypothesis

The Gaia (pronounced guy-ah) hypothesis is the idea that Earth is a living organism and can regulate its own environment. This idea argues that Earth is able to maintain conditions that are favorable for life to survive on it, and that it is the living things on Earth that give the planet this ability.

The idea that Earth and its atmosphere are some sort of "superorganism" was actually first proposed by the Scottish geologist (a person specializing in the study of the Earth) James Hutton (1726–1797), although this was not one of his more accepted and popular ideas. As a result, no one really pursued this notion until some two hundred years later, when the English chemist, James Lovelock (1919–), put forward a similar idea in his 1979 book, *Gaia: A New Look at Life on Earth.* Gaia is the name of the Greek goddess of the Earth, and in modern times has come to symbolize "Earth Mother" or "Living Earth." In this book, Lovelock proposed that Earth's biosphere (all the parts of Earth that make up the living world) acts as a single living system that if left alone, can regulate itself.

Lovelock arrived at this hypothesis (theory) by studying Earth's neighboring planets, Mars and Venus. Suggesting that chemistry and physics seemed to argue that these barren and hostile planets should have an atmosphere just like that of Earth, Lovelock stated that Earth's atmosphere is different because it has life on it. Both Mars and Venus have an atmosphere with about 95 percent carbon dioxide (gas), while Earth's is about 79 percent nitrogen (gas) and 21 percent oxygen. He explained this dramatic difference by saying that Earth's atmosphere was probably very much like that of its neighbors at first, and that it was a world with hardly any life on it. The only form that did exist was what many consider to be the first forms of life—anaerobic bacteria that lived in the ocean. This type of bacteria cannot live in an oxygen environment, and its only job is to convert nitrates to nitrogen gas. This accounts for the beginnings of a nitrogen buildup in Earth's atmosphere.

The oxygen essential to life as we know it did not start to accumulate in the atmosphere until organisms that were capable of photosynthesis evolved. Photosynthesis is the process that some algae and all plants use to chemically convert the Sun's light into food. This process uses carbon dioxide and water to make energy-packed glucose (sugar), and it gives off oxygen as a by-product. These very first photosynthesizers were a blue-green algae called cyanobacteria that live in water. Eventually, these organisms produced so much oxygen that they put the older anaerobic bacteria out of business. As a result, the only place that anaerobic bacteria could survive was on the deep-sea floor (as well as in heavily waterlogged soil and in our own intestines). Lovelock's basic point was that the existence of life (bacteria) eventually made the Earth a very different place by giving it an atmosphere.

Lovelock eventually went beyond the notion that life can change the environment, and proposed the controversial Gaia hypothesis. He said that Gaia is the "living Earth" and that Earth itself should be viewed as being alive. Like any living thing, Earth always strives to maintain homeostasis (constant, or stable, conditions) for itself. In the Gaia hypothesis, it is the presence and activities of life that keeps Earth in homeostasis and allows it to regulate its systems and maintain steady-state conditions.

Lovelock claims that it is the living things on Earth that provide it with the feedback so necessary to regulating something. (A feedback mechanism is something that can detect and reverse any unwanted changes.) Lovelock offers several examples of cycles in the environment that work to keep things on an even keel. He also warns that since Earth has the capacity to keep things in a stable range, human tampering with Earth's environmental balancing mechanisms places everyone at great risk. While environmentalists insist that human activity is upsetting Earth's ability to regulate itself, others argue that Earth can continue to survive very well no matter what humans do exactly because of its built-in adaptability. An important aspect of the Gaia hypothesis is that it offers scientists a new model to consider. It places great emphasis on what promises to be the planet's greatest future problem—the quality of Earth's environment.

Gene

The gene is the basic unit of heredity. All living organisms have genes that are the part of the cell that determines the traits offspring inherit from their parents. Genes are composed of deoxyribonucleic acid (DNA) and form part of the cell's chromosome. Genes produce their individual effect chemically by issuing the necessary instructions that tell a cell to make certain proteins.

Genes have been described as recipes for making proteins. Proteins build and control the activities of cells, so by making different proteins at different times, genes act as switches that control and change the way cells work. There are approximately 80,000 genes in each molecule of DNA in the human body, and a single gene can be thought of as a section of a single strand of DNA. Thus, if each gene is a recipe, the DNA is the chemical language in which the recipes are written. Genes may be

Strands of DNA found in a gene. DNA is the genetic material that carries the code for all living things. (Reproduced by permission of The Stock Market. Photography by Howard Sochurek.)

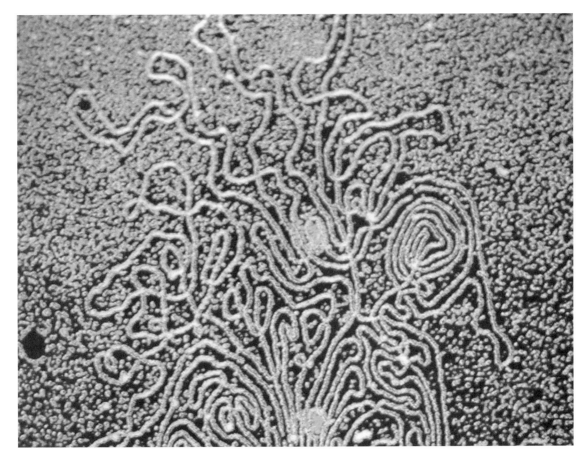

considered a message and DNA as the letters that make up the words of the message. Since DNA itself is made up of only four different chemicals called nucleotides—adenine (A), thymine (T), cytosine (C), and guanine (G)—all gene recipes or messages are also made up of some combination of these four "bases."

It was the Austrian monk, Gregor Johann Mendel (1822–1884), who introduced the world to the notion that there were certain factors in cells that determined all the specific hereditary traits. Even though Mendel never saw them or gave them their name, he knew by his experiments with cross-breeding peas that something like genes—tiny, independent units that passed on certain traits—existed. In 1910 the American geneticist Thomas Hunt Morgan (1866–1945) discovered that genes are located on chromosomes. Chromosomes were soon seen as threadlike structures that were made up of genes. Chromosomes were the chain, and genes were the links in the chain. Since each gene carries a different piece of information, each chromosome carries many different pieces of information. In the cells of the body, chromosomes exist in pairs, The same trait (such as height) is carried on both chromosomes of a certain pair, and it is always found in the same position or location on each chromosome. The two chromosomes of the pair carry the genes for either the same form (tall/tall) or the different form (tall/short) of a certain trait (height). These genes are passed from parents to their offspring when sex cells (sperm and egg) join during fertilization. The new cell formed by this union contains twenty-three chromosomes from the male and twenty-three from the female. Thus a new organism gets half of its genes (and therefore its traits) from each parent. After the American biochemist, James Dewey Watson, and his colleague, the English biochemist Francis Harry Compton Crick, discovered the molecular structure of DNA in 1953, it was soon learned how DNA duplicated itself and how it was able to form a specific protein. DNA was then proven to be the letters that formed the genetic code or message (which was the genes). Understanding genes and the role they play in life has transformed research in the life sciences, and has led not only to attempts to cure inherited diseases (known as gene therapy), but to a continuing push to understand the biochemical mechanisms in the body, especially the key proteins that govern all living processes.

[See also **Chromosome; DNA; Gene Therapy; Gene Theory; Genetic Code; Genetic Disorders; Genetic Engineering; Genetics; Inherited Traits; Mendelian Laws of Inheritance; Nucleic Acid**]

Gene Theory

Gene theory is the idea that genes are the basic units in which characteristics are passed from one generation to the next. Genes themselves are the basic units of heredity. The gene theory provides the basis for understanding how genes enable parents to transmit traits to their offspring. It is also a key element in the study of genetics.

Genes are the central objects studied by the science of genetics. The theory of genes (or gene theory) enables the science of genetics to be able to explain how information that is needed to make a new organism is passed from one generation to the next. Today we know that genes are made up of deoxyribonucleic acid (DNA), and we are able to state clearly what are known as the rules or laws of inheritance. However, less than 150 years ago scientists knew nothing about what went on at the cellular level that affected heredity. Since then, the science of heredity, or genetics (taken from the Greek word *genes* meaning "born") has been making regular and spectacular advances, so that at the beginning of the twenty-first century, scientists are close to learning the entire set of genetic instructions that go to form a single human being. Today scientists know that the 80,000 or so genes that make up what might be called the human blueprint are so individual that no two people (in a world of billions) are exactly alike—except for identical twins. Gene theory shows us how this extreme individuality can actually occur.

GREGOR MENDEL DISCOVERS DOMINANT AND RECESSIVE TRAITS

The existence of something like genes was recognized by the Austrian monk Gregor Johann Mendel (1822–1884), whose experiments with breeding different types of pea plants led him to describe what he called "hereditary factors," or genes. The first thing that Mendel discovered was that in crossing plants with different pure traits, such as all-tall plants with all-short ones, only a single trait was expressed. He therefore considered this expressed trait "dominant." Traits were therefore not blended, resulting in a medium-height plant, but were "expressed" as individual traits. He also found that a regular ratio of 3 to 1 existed for the number of dominant (tall) versus recessive (short) traits. This led him to decide that plants must contain what he called "factors" and "particles of inheritance," or what is now called genes. Mendel's other contribution was his correct assumption that both male and female parent contributed one "factor" per trait to an offspring. By 1900 it was realized that Mendel had given biology the basis for a new science of heredity, and the search began for the single "factor," or key, in all living things that contained the crucial information that dictated every detail of what an organism would be.

WATSON AND CRICK DISCOVER DNA STRUCTURE

By 1900, the existence of chromosomes was also known, and three years later, the American geneticist Thomas Hunt Morgan (1866–1945) announced the findings of his fruit-fly experiments, stating that chromosomes (the coiled structure in a cell that carries the cell's DNA) were made up of other, smaller things—later called genes. It was not until 1953 that the American biochemist, James Dewey Watson (1928–), and his colleague, the English biochemist, Francis Harry Compton Crick (1916–), were able to explain the molecular structure of DNA. With this new understanding, life scientists could formulate a fuller and more satisfying gene theory. Put simply, chromosomes are found in nearly every cell of our bodies. Chromosomes are made of DNA, and DNA stores genes. It is genes that carry the vital codes and information that not only tell a cell what to do, but which get passed on to the next generation by sexual reproduction.

The final part of gene theory explains how traits are passed on, and how no two individuals are exactly alike. During sexual reproduction, when a single human sperm fertilizes a single human egg, each contains only half the full set of forty-six chromosomes. Unlike other cells in the human body that have a complete set of forty-six chromosomes, sex cells contain only twenty-three. Consequently, when egg and sperm unite, the first new cell created gets twenty-three chromosomes from the mother and twenty-three chromosomes from the father to form a complete set of forty-six. This process, along with other "shuffling" of genes that occurs, guarantees that the new organism created is a unique individual.

Gene theory is the key to the genetics of the twenty-first century. Understanding how genes work and the knowledge that genes can change, or mutate, will lead to the prevention and cure of genetic diseases, as well as to the use of genetic engineering (the deliberate alteration of a living thing's genetic material to change its characteristics) to improve certain animal and plant species.

[*See also* **Chromosome; DNA; Gene Therapy; Genetic Code; Genetic Disorders; Genetic Engineering; Genetics; Inherited Traits; Nucleic Acid**]

Gene Therapy

Gene therapy is the process of manipulating genetic material either to treat a disease or to change a physical characteristic. It is accomplished by introducing a normal gene into the cell to make up for a defective or missing gene. Although some successes have been achieved, this technique is still in the research stage.

Once the scientific knowledge about genes and genetics had reached a certain level of sophistication by the early 1970s, scientists started thinking seriously about going right to the source of certain conditions and diseases and replacing bad genes with good genes. That is how gene therapy might be described in a simple way. Instead of introducing a therapeutic product like a drug into the body, scientists would try to deliver a gene to cure the problem once and for all.

In theory, the notion of gene therapy seems very attractive. It would allow new genetic material to be inserted into the cells in a patient's body to correct or improve a particular cell function; to make diseased cells even more vulnerable to being destroyed; or to block the functioning of diseased cells altogether. Again in theory, the way this is accomplished sounds feasible. If an individual has a disease that is the result of inher-

A scientist using computerized equipment to perform a DNA microinjection, a form of gene therapy. (Reproduced by permission of Photo Researchers, Inc.)

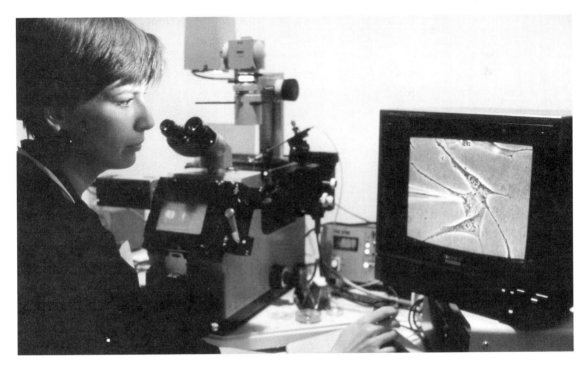

iting two copies of a defective gene, he or she would be a candidate for gene therapy. First, a new and properly functioning gene would be isolated from normal cells. Called the "transfer gene," this gene would be removed by cutting deoxyribonucleic acid (DNA) at specific locations using a technique called "gene splicing." Then a genetically altered virus would be used as the vehicle (called a "vector") to deliver the new gene to the cells. In fact, this has already been accomplished, and the first successful gene therapy experiment was performed in 1990. In that case, doctors replaced the defective genes of a four-year-old girl whose immune system did not work. Her body would not produce the necessary enzyme named adenosine deaminase (ADA). Doctors inserted a normal ADA gene into immune cells taken from her body and then returned the treated cells to her with a blood transfusion. The new gene gave the cell instructions to produce the enzyme and her immune system began to recover.

STILL NEEDS TO BE PERFECTED

Despite this success, researchers have discovered that the major limitation of gene therapy is the delivery system. They must always use a safe, efficient vector, or vehicle, to carry the new DNA to targeted cells. The early use of viruses as vectors was a logical strategy since viruses have been easily invading people's cells and causing trouble for thousands of years. Once there, viruses inject their own genes into the human cells, and the viral cells instruct them to make more viruses. Biologists have tried to turn the tables on viruses, and they carefully treat viral DNA in the laboratory so that it cannot reproduce itself once it gets into the body. Then human genes are inserted into the treated viral DNA, and this modified viral DNA is then introduced into human cells. Because it is carried by a virus, the modified DNA quickly finds its way into the DNA of the cells and becomes active. In practice, however, viruses infect cells in an unpredictable manner and can sometimes cause unwanted results. For example, inserting viral DNA into a cell might accidentally trigger the cell's oncogenes—specialized genes that can activate an uncontrolled cellular growth of cancer. In 1993 gene therapy for cystic fibrosis had to be stopped when the patient developed lung inflammation. In 1999 a young male patient actually died from a gene therapy treatment that went wrong. In the fall of that year, eighteen-year-old Jesse Gelsinger of Arizona died after undergoing experimental genetic therapy a the University of Pennsylvania. The young man suffered from an inherited liver disorder and was given new genes as an attempt to cure him. While it is not yet known if mistakes were made in his treatment and therapy, his tragic loss underscores the dangers or conducting experiments on humans.

One alternative to the use of viruses in gene therapy is the technique of packaging genes into fatty globules. This has been used to fight tumors that often grow when the immune system fails to recognize the tumor cells as foreign and does not kill them. By wrapping the new gene in droplets of fat, called liposomes, which bind easily to cell surfaces, scientists have been able to experimentally insert a gene that marks the tumor cells in a way that forces the immune system to attack them. Researchers are working steadily at trying to make their methods of transferring genes into human cells more effective, efficient, and safe.

Although gene therapy is still in its early stages, researchers are confident that it will become a standard, accepted strategy for fighting diseases like cancer in the future. Contrary to what many people believe, however, gene therapy is different from genetic engineering since it does not target sex cells that pass along genetic traits. Rather, gene therapy attempts to fix existing problems in ordinary cells and does not tamper with genes in eggs or sperm. Thus, a person who has a single-gene disorder corrected by gene therapy can still pass on a faulty copy of a gene to his or her children. Unlike gene therapy, the more ambitious and fundamentally different goal of genetic engineering is preventative. Genetic engineering seeks to keep diseases from happening in the first place. This process is more controversial since it involves changing genes in the sex cells and gives rise to many ethical and legal issues.

[*See also* **Chromosome; DNA; Gene; Gene Theory; Genetic Code; Genetic Disorders; Genetic Engineering; Genetics; Inherited Traits; Nucleic Acid**]

Genetic Code

The genetic code tells a cell how to interpret the chemical information stored inside deoxyribonucleic acid (DNA). This information is in the form of a sequence of chemicals that tell a cell which proteins to make. Without the genetic code, the cell would be unable to interpret the DNA sequence, and therefore could not make the proteins that build cells and make them work.

By the early 1950s, scientists knew that genes were made of DNA, and that specific proteins were made by specific genes. DNA is found in the chromosomes in the nucleus of cells, and it controls the characteristics of living things by means of a chemical code of instructions. The structure of the DNA molecule was found to resemble a twisted ladder called a double helix. The rungs on this ladder are called "bases" and are the coded instructions. These instructions are written with only four chemicals—adenine (A), thymine (T), guanine (G), and cytosine (C)—that make up what might be considered a four-letter alphabet. These bases must pair up a certain way (A only with T, and G only with C). Each pair of bases is called a nucleotide. It is the order of these nucleotides along the DNA that spells out the instructions for making proteins, which control the characteristics of organisms.

Proteins are chains made of twenty different units called amino acids, and it is the order of the amino acids that determines what type of protein will be produced. After the 1950s discovery of the molecular structure of DNA, the question that drove geneticists during the 1960s was: "How is a gene, whose information is contained in the sequence of only a four-letter alphabet (A, T, G, C) able to code enough messages for twenty different amino acids?" If a single base coded for one amino acid, only four amino acids could be made. If two bases coded for one amino acid, then a maximum of sixteen arrangements was possible. However, if the four bases somehow combined in groups of three to form one amino acid, sixty-four combinations were possible. After a great deal of difficult research, this triplet code called a "codon," proved to be the answer. The explanation somewhat resembles that of the Morse code, which is able to code all twenty-six letters of the alphabet using only two symbols—a dot and a dash. It does this by using different combinations of dots or dashes to code for each letter of the alphabet. With DNA, the answer lies in the codons or triplet code. Each codon is three bases long and has an exact meaning. In other words, a group of three bases in a certain order forms the codon for a specific amino acid. Therefore, the sequence GAG would specify the amino acid glutamic acid.

Once the idea of a triplet code was discovered, years of work resulted in what might be called a working dictionary of codes. It was found that of the sixty-four possible combinations, sixty-one of the codons are actually used to form the twenty amino acids. This means that some amino acids can be specified by more than one type of codon. It also means that the remaining three codons do not code for any amino acid but instead act as punctuation in a long message. Thus, these three codons can signal the end of a genetic "sentence" and therefore the completion of a code. It makes sense that, just as a paragraph of words has punctuation guiding the reader, a continuous sequence of hundreds of thousands of bases needs punctuation to make it a meaningful set of precise instructions. These three codons therefore not only end a code, but are thought to also signal something like, "I am not a gene," or "I am not a gene but one is coming soon." Interestingly, however, no commas or internal punctuation are found within the code. Punctuation is limited to stop or start signals at the beginning or end of a continuous run of triplets. The code was also found to be linear, meaning that just like a sentence, its sequence of bases are supposed to be read from a fixed starting point (the beginning). Once this genetic code was broken, it was found that the code is universal. That is, the very same three-letter codons specify the exact same proteins for all living things—from humans to bacteria. All life is therefore guided or directed by a common language that is the genetic code written in all DNA.

[*See also* **DNA; Gene; Gene Theory; Genetics**]

Genetic Disorders

Genetic disorders are conditions that have some origin in a person's genetic makeup. Genetic disorders are more severe than simple abnormalities and usually result in some type of medical problem. A genetic disorder is not the same thing as a disease.

Genetic disorders are generally one of two types: those that are inherited (and are governed by the same rules that determine all our traits), and those that are the result of some type of mutation or change that took place while the embryo was developing. Sometimes this mutation is caused by environmental factors. Within this first set of inherited disorders, a disorder that is transmitted by genes inherited from only one parent is called an "autosomal dominant disorder." The term "autosomal" refers to any of the twenty-two sets of chromosomes (forty-four individual chromosomes) common to both males and females that determine all of the traits except a person's gender. (The single pair of chromosomes that determine sex is called the sex chromosome and is different between men and women.)

In an autosomal dominant disorder (ADD), an individual inherits a *dominant* gene from one parent that causes the disorder. Since it is dominant, the gene has a fifty percent chance of being expressed. Sometimes an individual will inherit a gene for a genetic disorder that is not a dominant gene. In that case, it is a recessive gene that is "covered" by the normal gene (since genes are always inherited in pairs), and the person will not have the disorder. There are approximately 2,000 ADDs, ranging from the inconvenient (like an extra toe) to eventual death (as in Huntington's disorder). Although certain disorders reduce a person's chance of surviving (and therefore of passing the gene on to offspring), many ADDs do not affect reproduction or are diagnosed fairly late in life.

Autosomal recessive disorders, also called recessive genetic disorders (RGD), are the result of both parents supplying a recessive gene to their offspring. In this case, the offspring has no chance to avoid the disorder since it does not have any normal gene to possibly "cover" for the recessive one. Instead, the individual has two recessive genes for the same trait. There are about 1,000 RGDs, and the odds for getting two of the same recessive genes are as high as 25 percent. Some of the better-known recessive genetic disorders are sickle-cell anemia, cystic fibrosis, and Tay-Sachs disease. Galactosemia is an example of what is called a metabolic RGD. In this case, a person with this disorder lacks a certain enzyme needed to break down the sugar found in milk. Another disorder, adenosine deaminase deficiency, is, like galactosemia, one of the few treatable genetic diseases. This immune disease is sometimes cured by a bone marrow transplant.

Certain other traits are not received from the set of twenty-two pairs of autosomal chromosomes but are instead found on a single pair of sex

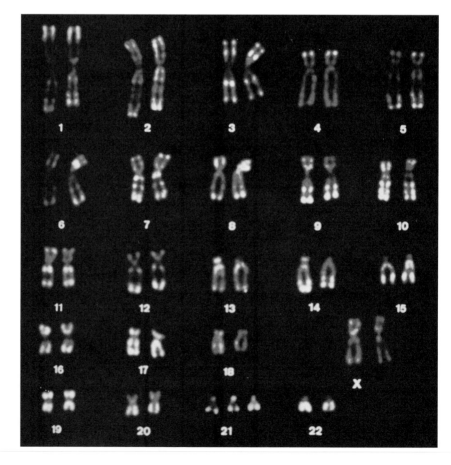

Down syndrome is a genetic disorder resulting from chromosome pair twenty-one receiving three chromosomes instead of just two. (Reproduced by permission of Custom Medical Stock Photo, Inc.)

chromosomes. Since females have two X chromosomes and males have one X and one Y, males are more susceptible to X-related gene defects. Since females have two X chromosomes, any defect on one is no more likely to be expressed than a gene defect on any other chromosome (unless they inherited two X's with defects). However, since males have only a single X chromosome, *any* defective gene on that chromosome will be expressed. The bleeding condition Hemophilia A is a sex-linked or X-linked recessive disorder, as is color blindness.

Finally, there are genetic disorders that are caused not by gene mutations but by things going wrong right from the beginning of fertilization and cell splitting. Sometimes an abnormal number of chromosomes are formed (either too few or too many), resulting in such conditions as Down syndrome in which children are born with some degree of mental retardation and sometimes heart defects. Although genetic disorders are not common, it is estimated that of all newborns, as many as one percent will have some form of chromosome-related condition.

[*See also* **Chromosome; Embryo; Gene; Gene Therapy; Gene Theory; Genetic Code; Genetic Engineering; Genetics; Mendelian Laws of Inheritance; Mutation; Nucleic Acid**]

Genetic Engineering

Genetic engineering is the deliberate alteration of a living thing's genetic material to change its characteristics. It is also a general term that describes a range of techniques that allow geneticists to transfer genes from one organism to another. The applications of genetic engineering are vast, and since it is technically possible to produce new gene combinations that could never occur in nature, the implications of this new technology are controversial.

Genetic engineering has actually been practiced under another name for thousands of years. Probably the oldest version of it was conducted under the name of agriculture when farmers deliberately crossed plants with certain desirable characteristics and did not breed those without them. They did the same thing with farm animals and called it selective breeding. For example, animal offspring that showed certain desired characteristics were bred with their like, while those not showing desired characteristics were not allowed to reproduce. It was such a technique that gave us the many different types of dogs we have today. However, selective breeding is a slow, trial-and-error process since it must allow animals the time to grow and mature sexually. By the 1970s, once science had developed ways to isolate individual genes and to reintroduce them into cells, they were able to directly and quickly alter the deoxyribonucleic acid (DNA) of an organism and accomplish overnight what would have taken generations.

RECOMBINANT DNA TECHNOLOGY

Genetic engineering is also called "recombinant DNA technology" because the DNA of existing organisms is actually recombined into new organisms. Genetic engineering is a form of gene manipulation that results in a new arrangement of genes. This is achieved by removing a certain part of DNA and attaching it to another piece of DNA, sometimes from a different organism or even a different species. According to what it was intended to do, this transfer might give its new host a new trait, or it might enable it to produce substances that it never before could. Certain crops have been genetically altered or engineered to withstand the

DNA being injected into a mouse embryo. The discovery of DNA in 1953 led to the new field of genetic engineering. (Reproduced by permission of Archive Photos, Inc. Photograph by Jon Gordon.)

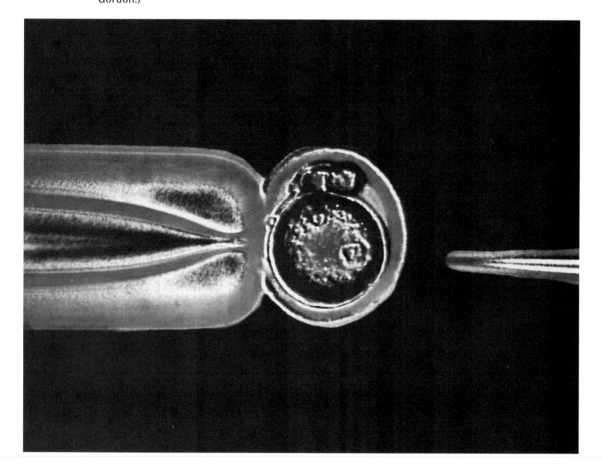

effects of a herbicide that kills weeds. Other plants have had their ability to resist certain diseases transferred genetically to plants without that natural ability. In medicine, genetic engineers have manipulated bacteria and made them cheap, efficient producers of substances needed by humans.

Most of today's genetic engineering is accomplished by inserting genes into bacteria. This recombining is achieved by a relatively simple process described as "cutting and pasting." First a gene is identified, isolated, and spliced or cut out of its DNA strand. In gene splicing, special proteins called "restriction enzymes" are used as scissors. These enzymes react chemically with a certain part of the DNA and break it off, leaving each piece with what are described as "sticky ends." With the help of another enzyme called ligase, these ends will easily attach to any other piece of DNA, even if it comes from a different organism. It is because of this ability that DNA is so easily transferrable between organisms, even ones as different as humans and bacteria. After the gene-containing DNA has been cut out and spliced to another DNA fragment, a hybrid (having mixed composition) molecule called recombinant DNA is formed and inserted back into the cell. When the cell divides, the number of recombinant DNA molecules also increases.

GENETIC ENGINEERING HAS MANY USES

Today genetic engineering has found many uses in agriculture, industry, and medicine. Plants have been engineered to withstand herbicides that kill weeds, as well as to resist insects and even grow in poor soil. In one of the more amazing experimental uses of genetic engineering, genes from fish that thrive in arctic waters have been spliced into plants in an attempt to make them tolerate freezing temperatures. A genetically altered tomato, the "Flavr Savr," can remain on the vine longer and thus does not have to be picked and shipped while still green. In industry, genetic engineering is creating microorganisms that can more efficiently clean up oil spills by naturally breaking down the oil. This also has potential for neutralizing toxic substances and other waste products.

Genetic engineering has received most of its public notice in the medical field. Already, bacteria have been genetically engineered to produce human insulin in large quantities, enabling diabetes treatment to be easier and less expensive. The same has been done for the production of human hormones. Gene therapy is a form of genetic engineering, which introduces a normal gene to make up for a missing or defective gene. Finally, for those who might pass a genetic disease on to their offspring, genetic counseling is now available that allows people to be more informed before they decide to have children.

The recently acquired ability to alter organisms at the level of their genes is more than just another powerful tool of modern science. If we can insert a normal gene in a human to replace a defective one, who is to say we cannot change or modify an existing trait in order to have one that is more "desirable?" The ability to manipulate human beings genetically is an awesome responsibility that some argue should never be attempted. Others say it should be done only under strict guidelines, while others are simply confused by its far-reaching social and ethical implications. There is always the fear that an accident may occur and an uncontrollable bacteria or some other engineered life form might prove to be environmentally disastrous. However, genetic engineering or recombinant DNA technology has such enormous potential for useful and beneficial applications in so many areas that it can never be simply ignored. Fortunately, ethical and legal committees from many disciplines are in place to oversee this crucial work and to consider its future.

[*See also* **Chromosome; DNA; Gene; Gene Therapy; Gene Theory; Genetic Code; Genetic Disorders; Genetics; Inherited Traits; Nucleic Acid**]

Genetics

Genetics is the branch of biology that is concerned with the study of heredity or the passing on of characteristics from one generation to the next. It is also concerned with variation or what makes one living thing different from another. Geneticists are people who study genes in an attempt to understand how the inherited information genes contain is stored and passed on.

Genetics is a fairly young science and was started by the landmark work of the Austrian monk and botanist Gregor Johann Mendel (1822–1884), who first put forward his theory of heredity in 1865. Until Mendel, most everyone knew that noticeable traits were usually passed on from one generation to the next, but no one knew where to begin to find out what controlled and influenced the passing on of these traits. It was Mendel who carried out the first scientific study of how traits pass from one generation to the next by conducting a series of experiments with pea plants. The experiments took him eight years to complete. Mendel decided to use ordinary green peas, like the ones commonly eaten today, since they are easy to breed for what are called "pure traits." This means that a purebred plant that produces yellow pods will always produce yellow pods. Mendel selected pea varieties that differed in a single trait (such as height or color or wrinkled seeds) and then crossed them with plants of a different trait (short with tall, yellow with green). For example, he would cross a pure tall with a pure short and then accurately record the number of each type he harvested for each generation. Luckily for Mendel, he had concentrated on only a few traits, and each of these happened to be governed by a single gene (which he did not yet even know existed). The first thing that Mendel discovered was that crossing a tall plant with a short one did not result in a blend of medium-height plants, but rather in plants that were all tall. After allowing these offspring plants to pollinate (the transfer of pollen containing male sex cells to the pistil containing female sex cells) themselves, he found that the next generation produced plants three-quarters of which were tall (which he then called a dominant factor) and one-quarter of which were short (which he called a recessive factor). After crossing hundreds of plants and keeping careful records, Mendel discovered that a regular 3 to 1 ratio or pattern existed for the number of dominant versus recessive traits. This led him to decide that plants must contain what he called "factors" and "particles of inheritance," or what we now call genes. This also led him to believe that there must be laws or rules that determined how these "particles" were passed on. After much study, Mendel formulated what are now called the "Mendelian laws of inheritance." He stated correctly that traits did not blend but remained distinct; that they combined and sorted themselves out according to fixed rules; and that both male and female contributed equally. Mendel was also correct when he said that each parent contributed one "factor" for a particular trait to its offspring. Although Mendel's laws had laid the foundation of the new science of genetics, his work was unknown until 1900 when it was separately discovered by three different botanists (in three different countries) who realized that Mendel had discovered the laws of genetics long before they did. Although each published his own version of these laws, each cited Mendel as having been the real discoverer, and all said that their work was merely a confirmation of what Mendel had accomplished thirty-five years before.

MORGAN INTRODUCES CHROMOSOME THEORY OF INHERITANCE

By 1900, when biologists realized that Mendel had given them the basis of a new science of heredity, they began to search for the key part in all living things that contained the crucial information that determined every detail of what an organism would look like. This search soon led to an understanding of chromosomes and then to the discovery of a single molecule called deoxyribonucleic acid or DNA. By 1900, science already knew that chromosomes (the coiled structure in a cell that carries the cell's DNA) existed in the nucleus of every cell, and by 1903, some biologists were making a connection between chromosomes and the process of heredity. One scientist, the American geneticist Thomas Hunt Morgan (1866–1945), conducted experiments on the rapidly bred fruit fly and arrived at the idea that chromosomes are themselves made up of other, smaller things (later named "genes") that were linked together and arranged in a long line. In 1911 he put forth this chromosome theory of inheritance based on genes. The next question was "how did the genes actually work?"

As early as the late nineteenth century, it was known that a unique type of acid named deoxyribonucleic acid (DNA) existed in every cell nucleus. It had been long ignored, however, and was considered unimportant to heredity. By 1944, though, better research and improved technology enabled biologists to realize that DNA was actually the key chemical at the center of heredity. Research continued and by 1952, biologists were able to demonstrate that DNA was indeed the genetic material for which they had been searching. However, since its structure was still unknown, it was not possible to describe how such an apparently simple molecule of DNA could contain the vast and complicated information or code that was needed to develop a human being.

WATSON AND CRICK COMPLETE THE PUZZLE

The final breakthrough to this puzzle was achieved in 1953 by the American biochemist James Watson and his colleague, the English biochemist Francis Harry Compton Crick. That year they were able to successfully construct a "double helix" model of a DNA molecule that solved the puzzle. DNA, they explained, is made up of two long strands connected by "base pairs" (like rungs on a ladder), and that the entire model looks like a curving, twisting ladder or a spiral staircase. Watson and Crick also found that the bases always paired up in a specific order, so that if they knew the sequence of one strand, they could accurately tell the sequence of the other. Finally, they discovered that the order of the chemical bases represented a code that was translated by the cell and used as a guide to make proteins.

Since their achievement, scientists have learned that DNA is the blueprint for all life on Earth. It is now known that almost every cell in our bodies has a set of chromosomes that store this DNA. It also is known that each DNA base is like a letter in the alphabet, and that a sequence of bases forms a message. These messages are called genes, and each gene instructs the cell that contains it on how to make a specific protein.

We also know that while all living things have their own unique code of DNA, the molecule that forms that code is basically the same for all.

Finally, we know that it is through chromosomes that genetic information is passed from one generation to another. Today's genetics has gone beyond all of these discoveries and has begun to put them to good use. Now that scientists actually are able to look at the genes themselves, they can know what sequences produce what effect. This sometimes allows scientists to move genes around, fix mistakes, and even transfer them between species. These many years of research culminated in 1997, when scientists in Scotland produced a lamb that was cloned from a cell nucleus taken from the udder of a sheep.

As the twenty-first century begins, science is preparing to cope with what promises to be a difficult and probably very controversial period as it becomes able to artificially alter genes through a process called genetic engineering. This powerful new tool can produce great benefits, such as curing a genetic disease like Alzheimer's, but it also has the potential for serious misuse. Genetics may be a young science, but it promises to put its stamp on the twenty-first century.

[*See also* **Chromosome; DNA; Gene Therapy; Gene Theory; Genetic Code; Genetic Disorders; Genetic Engineering; Genetics; Inherited Traits; Nucleic Acid**]

Genus

The term genus is one of the seven major classification groups that biologists use to identify and categorize living things. These seven groups are hierarchical or range in order of size, and genus is one of the smaller, important, and more frequently used groups. The classification scheme for all living things is: kingdom, phylum, class, order, family, genus, and species.

Coming as it does between the larger group, family, and the smaller group, species, members of the same genus have more in common than those in the same family and less than those in the same species. Although members of the same genus are very similar (like a wolf and a coyote), members of different groups usually cannot breed with one another. Members of the same genus, however, are known to be very closely related in terms of their evolutionary history, and it is obvious that they share the same basic shape and structure as well as similar biochemistry (the chemistry of biological processes) and even behavior.

The group genus is almost always used with the more particular grouping, species. All organisms are referred to scientifically by a two-word, Latin name called a binomial. Humans therefore are *Homo sapiens*. This example of *Homo* is unusual, for only one living species occurs in that genus. Most genera are "polyspecific" and contain more than one species. This is especially the case for plants known as sedges (*Carex*) and insects known as fruit flies (*Drosophila*), each which has hundreds of species in the genus.

[*See also* **Class; Classification; Family; Genus; Kingdom; Order; Phylum; Species**]

Geologic Record

The geologic record is the history of Earth as recorded in the rocks that make up its crust. Rocks have been forming and wearing away since Earth first started to form, creating sediment that accumulates in layers of rock called strata. The way these strata are arranged and what fossils are in them give scientists clues about what Earth was like billions of years ago.

The concept of what is called geologic time is somewhat difficult to fully grasp because it deals in such enormous blocks of time. When people first began to seriously study Earth around the seventeenth century, their first estimate of Earth's age was in the thousands of years. One famous example is that of the Irish clergyman James Ussher (1581–1656), who used the Bible to calculate that Earth was created in 4004 B.C. A century later, estimates by others had only raised that number to about 75,000 years, and it was not until the Scottish geologist (a person specializing in the study of Earth) James Hutton (1726–1797) made his famous statement that the Earth contains "no vestige of a beginning—no prospect of an end," did the notion of millions and perhaps billions of years begin to be considered. Today, with advanced tools, scientists are able to say with some certainty that Earth is about 4,600,000,000 years old.

READING THE GEOLOGIC RECORD

By examining the progress of geology, humans touch upon the major breakthroughs that allowed scientists to be able to "read" Earth's geologic record. This geologic history of the planet's evolution (changes occurring over time) and developmental changes is recorded in its rocks. One of the earliest breakthroughs was the eighteenth-century realization that there was something to be learned by the obvious relationships of one type of rock to another. Called the law of superposition, this idea stated that in an undisturbed section of sedimentary rocks (formed in layers or "strata" by weathering and erosion), each layer is older than the one above it and younger than the one below it. Although this seems fairly simple and obvious today, it was a major breakthrough in being able to start to date the age of Earth.

A related idea that also proved very helpful was the principle of faunal succession. This states that the fossils found in rocks also succeed one another in a definite order, and that a time period can be recognized by the type of fossils contained in its rocks. This principle applies throughout the world, so that geologists can identify rocks of the same age even if they are found in widely separated locations. The importance of fossils cannot be overemphasized, since without them, science would lose its primary tool for subdividing geologic time periods into smaller and smaller chunks.

ERAS, PERIODS, AND EPOCHS

Geologists have divided the geologic record into periods that can be organized or charted onto a timescale. The major divisions of geologic time are known as eras which are described by some as "chapters" in Earth's history. Each era is naturally different from another, especially in terms of the nature of life it contained. The eras are then divided into periods. These have nothing to do with the passage of a certain amount of time, meaning that they are not of equal length. Instead, they are based upon the nature of the rocks and fossils found there. Some may be longer than others. The main subdivisions of periods are called epochs. These eras, periods, and epochs usually were named after places on Earth (mostly in Western Europe) where the rocks of those times were first discovered.

In terms of the geologic record, life on Earth is first seen about 3,500,000,000 years ago. This means that for about 1,000,000,000 years from the time Earth first formed, there was no life on the planet. The oldest primitive fossils found were simple prokaryotic organisms (bacteria whose cells did not have a nucleus or any other structures). The first eukaryotic organisms (whose cells contain a nucleus that is surrounded by a membrane) appeared about 1,800,000,000 years ago in what is called Precambrian era.

The first multicellular organisms did not appear on Earth until somewhere between 700,000,000 and 1,000,000,000 years ago. Then, during the Cambrian period of the Paleozoic era, an explosion of multicellular

CHARLES LYELL

Called the father of modern geology (the study of the Earth), Scottish geologist Charles Lyell (1797–1875) offered proof that Earth's surface was the result of natural forces operating very slowly over millions of years. His work laid the foundations not only for modern geology but for the study of evolutionary biology as well.

Charles Lyell was born to a well-to-do family in Kinnordy in eastern Scotland; his family moved to Southampton, England, when he was two years old. His father was an amateur naturalist who kept a well-stocked library that the young Lyell often used. As a youngster, Lyell was more interested in observing nature and collecting butterflies and insects than in school. He entered Oxford University at nineteen and his interest in geology increased, although he prepared for a law career. At twenty-two he graduated and moved to London to study law, although he spent every free moment on geological trips and excursions. By the time he finished law school and was admitted to the bar, he had conducted several geological tours in England and the continent and met with some of the best minds in geology.

By the age of twenty-eight, Lyell still had his father's financial support and was doing much more geology than law. By now, he was doing serious geological research and writing and beginning to formulate his own ideas. After making a long and difficult geological trip through France and Italy, Lyell returned to begin his *Principles of Geology,* which would become one of the most influential textbooks ever written. The state of geology at this time was such that it was still dominated by individuals who believed that Earth was not very old, perhaps only several thousand years, and that the obvious changes that had taken place were the result of sudden, catastrophic occurrences. Lyell's readings, and mostly his field trips, had convinced him of just the opposite. He agreed with the earlier ideas of his coun-

life in the sea took place. Continuing explosions occurred, going from marine invertebrates (animals without a backbone) to the beginnings of actual fishes. Around 435,000,000 years ago, the first land plants evolved, followed by great swamp trees, amphibians, and primitive reptiles. The dinosaurs came on the scene during the Triassic period (about 225,000,000 years ago), and by about 180,000,000 years ago, the Jurassic period saw the first birds and mammals.

Sometime during the Paleocene epoch of the Cenozoic era (about 65,000,000 years ago), the Age of Mammals began. Around 2,000,000 years ago, *Homo habilis* appeared and was the first human species to be given the genus name *Homo,* meaning "man."

tryman, the Scottish physician James Hutton (1726-1797), who had long before put forth his idea of "uniformitarianism." This theory (written before Lyell was born) said that natural geological forces, like earthquakes, volcanoes, and erosion acted upon the Earth over an extremely long period of time, and that geological change was very slow-acting and sometimes barely noticeable. Lyell's *Principles of Geology* was published in three volumes from 1830 to 1833. It was so well-written and so well-documented, that despite the fact that it contained few original ideas, it communicated and explained the basic principles of the new geology so well and so effectively that it became a steady best-seller. It went through twelve editions in Lyell's lifetime.

One of those who closely read Lyell's books and who was deeply influenced by them was the English naturalist Charles Darwin (1809-1882). Darwin and Lyell had become good friends, and when Darwin left to take his famous trip in 1831 on the H.M.S. *Beagle,* he had the first volume of Lyell's book with him. Lyell's ideas of timelessness and gradual change proved to be highly influential when Darwin began to formulate his own ideas about biological evolution (physical changes that occur over generations). In fact, Darwin is said to have once stated, "I always feel as if my books came half out of Lyell's brain." Darwin admitted that he drew heavily on Lyell's book, both for its excellent writing style and its real content. As a result, both men would shape the thinking of their disciplines with their ideas that Earth and life itself was much more ancient than thought, and that existing species appeared to have evolved from previous ones now extinct. Another important geological concept championed by Lyell concerned what is called the geological record. That is, that older rocks are generally buried beneath younger ones, and that careful excavation and study of the geological layers and fossils found there contain the evolutionary history of Earth itself.

Finally, because of the geologic record, scientists also know that a major phenomenon like continental drift (the movement of the plates on Earth's crust) is responsible for the current position of the continents. Scientists also know that the record contains evidence of several extinctions. None of these are more dramatic and puzzling than the disappearance of the dinosaurs about 100,000,000 years ago during the Cretaceous period. Scientists have described the geologic record as a history book to be read to learn about Earth's past. However, the geologic record is only useful if one knows how to read its signs and interpret them.

[*See also* **Fossils; Radioactive Dating**]

Germination

Germination is the earliest stages of growth when a seed begins to transform itself into a living plant that has roots, stems, and leaves. Although conditions vary according to species, all seeds require a certain amount of moisture and oxygen as well as a suitable temperature before they will germinate.

Some seeds are ready to germinate almost as soon as they are ripe and will sprout open wherever and whenever they land in a suitable environment. However, the seeds of most plants need to lie dormant (inactive or resting) for a period of time before they will germinate. This enforced dormancy can be caused by many factors within the seed itself. First, the seed coat may be so hard that it will not allow water or oxygen to penetrate until it has begun to soften or break down over time. Second, seeds may contain chemicals that prevent germination, and sprouting will not occur until these antigermination hormones have been washed away by rainwater. Other seeds need to be exposed to prolonged periods of cold, while others must pass through the gut of an animal or even be exposed to fire before germination will occur. During this dormancy period, the seed is inactive and no growth occurs. Seeds have remarkable properties, and some can remain dormant for extremely long periods of time. In fact, some seeds have been known to germinate after remaining dormant for centuries.

A mature or ripe seed is surrounded by a hard coat called a testa. Inside this coat is the beginning of a plant called an embryo. The embryo has one or more seed leaves called cotyledons. Also inside is all the food the embryo will need to fuel its early growth. Germination usually happens in the spring when the soil warms, and the seed breaks its dormancy. At the beginning of germi-

A kidney bean sprouting into a seedling. This is the beginning of germination, a process that cannot be reversed. (Reproduced by permission of Photo Researchers, Inc.)

nation, the seed takes in water very quickly. This process is called imbibition, and as the dry seed takes in more water, it swells and finally bursts the seed coat. Once dormancy ends and the seed coat bursts, germination becomes irreversible. With its coat now open, the seed can take in even more water and oxygen, so that it soon doubles in size.

The next stage of germination begins when the food stored in the endosperm is converted into useable forms and sent to the seed's growing points. These points consist of the seeds beginning root system called the radicle and its early stem and leaf stage called the plumule. The radicle or young root is the first to emerge from the seed and begins to grow downward into the soil. Soon after, the young shoot or plumule appears and starts to grow upward. The plumule breaks through the soil with its tip bent over, protecting the young, tender tip and allowing the older, stronger part of the shoot to bear the brunt of pushing upwards. In some plants like garden beans, the cotyledon (the first leaf to appear from a sprouting seed) is also raised out of the ground, while others, like peas, the cotyledon stays buried. After the radicle becomes a root system and the plumule straightens out, the cotyledon begins to open and the first true leaves start to grow. By now, the seedling is well established and the life cycle of another generation has begun.

[*See also* **Botany; Embryo; Fertilization; Seed**]

Golgi Body

A Golgi body is a collection of membranes inside a cell that packages and transports substances made by the cell. All eukaryotic cells have one or more Golgi bodies that work closely with the endoplasmic reticulum (a network of membranes, or tubes, in a cell through which materials move). A eukaryotic cell is one with a distinct nucleus (such as a plant or animal cell).

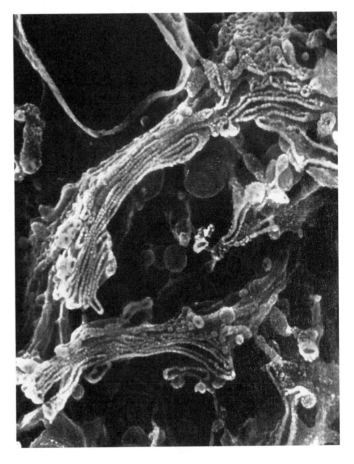

A high resolution scanning electron micrograph of the Golgi body of an olfactory bulb cell. The Golgi body collects proteins and lipids made by the endoplasmic reticulum. (©Photographer, Science Source/Photo Researchers, Inc.)

Golgi bodies were named after Italian anatomist (a person specializing in the structure of animals) Camillo Golgi (1843–1926), who first saw them in brain cells late in the nineteenth century. Golgi was able to see these tiny organelles (or cell structures that have certain functions) because of his discovery of a particular cell stain. Golgi found that by staining nerve cells with silver nitrate, he was able to see details that were not otherwise visible. He was also able to make out all of the very fine extensions that branched off the Golgi bodies. Looking like a stack of flattened bags, these piles of membranes are usually found close by the nucleus and work closely with the cell's endoplasmic reticulum (which make proteins). Their function is to package the proteins made by the cell, and they literally wrap these proteins in membranes. The proteins then travel either to another part of the cell or to the cell membrane to be transferred outside the cell. Golgi bodies also work on any "unfinished" proteins that the endoplasmic reticulum has created. Sometimes these proteins need a bit more tinkering or finishing before storage or use, and it is the Golgi bodies that perform whatever chemical modifications are necessary. Golgi bodies are also called Golgi apparatus. They have also been referred to as the Golgi complex.

[*See also* **Cell; Endoplasmic Reticulum; Membrane**]

Grasslands

Grasslands are particular geographic regions in which grasses are the dominant vegetation. Grasslands typically have only two seasons: a long, dry season and a shorter, rainy season. In their natural state, grasslands can support a variety of large grazing animals.

Most grasslands are called temperate grasslands since they spread across geographic areas where conditions are more moderate than extreme. In temperate grasslands, summers are hot and winters are cold. Grasslands receive less rain than forested areas but more rain than deserts. Grasslands are found in both the Northern and the Southern Hemisphere and are called by a variety of names. In North America, they are the "plains," in Argentina and Chile they are the "pampas," in the Andes of South America they are called "paramo," in Eurasia the "steppes," the "veld" in South Africa, and the "outback" in Australia. Besides sharing similar climates, all of these regions have few trees, except near rivers and streams. The land is either flat or has rolling hills. Different types of grasses grow in different regions, although all have adapted to moist winters and dry summers. Some grass roots may go as deep as 9 feet (2.74

meters) in search of water, although more often they rely on a huge, spreading system of roots and subsurface stems called rhizomes.

In the United States, what are called shortgrass prairies existed east of the Rocky Mountains. These were home to some 60,000,000 bison at one time. Shortgrass prairies, such as those found in South Dakota, experience strong winds, extreme light, and infrequent rainfall. Most of the original North American Great Plains have been plowed under to grow wheat. This contributed to the Great Dust Bowl of the 1930s when poor farming practices, drought, and strong winds transformed the prairie into a wasteland.

Tallgrass prairies also once extended west across North America from the temperate deciduous (trees that lose their leaves before winter) forests. It is said that they grew as high as a person on horseback. These grasslands were covered with herds of grazing animals, especially bison. The main predators were wolves, although coyotes predominated in the drier

Stands of trees in a grassland during the summer. (Reproduced by permission of Field Mark Publications. Photograph by Robert J. Huffman.)

west. Smaller animals included prairie dogs and their predators (foxes, ferrets, hawks, and eagles). In the American Midwest and the Russian steppe, most wild herbivores (animals that eat mainly plants) have all but disappeared. In the grasslands of Africa, buffalo, antelope, and rhinoceros all live together. In African tallgrass, a particular type of "grazing succession" takes place where each herbivore has a niche or proper role to play. As the zebras eat the tall grass, they expose the lower grasses. These grasses are grazed on by the wildebeest whose teeth cannot handle the tall grass. Wildebeest grazing simulates the growth of new grasses, which is then consumed by Thomson's gazelle. Many of these animals eat grass seeds that are then passed through their systems and deposited elsewhere in their dung (waste).

Grasslands have proven well suited to farming, since they have few trees, rich soil, mostly flat, that little of any original grasslands now exist. The temperate grasslands have become what is called the "breadbasket of the world." Corn and wheat have mostly replaced grasses. However, ecologists are concerned that this switch from grasses to crops has increased the amount of carbon dioxide in the atmosphere, possibly contributing to global warming.

[*See also* **Biome**]

Greenhouse Effect

The greenhouse effect is the name given to the trapping of heat in the lower atmosphere and the warming of Earth's surface that results. Although a natural phenomenon, this warming effect tends to increase as more human-produced gases are released into the atmosphere. This increased warming could result in climate changes that affect crops, as well as melting glaciers and coastal flooding.

Only in the past few decades has the term "greenhouse effect" implied something bad for the environment. In fact, the phenomenon that it describes is absolutely essential to life on Earth. Without it, Earth would be a cold and lifeless planet. What this phenomenon does is hold on to some of the heat given off by the Sun. Specifically, Earth's atmosphere is the mixture of gases and water vapor that surrounds the entire Earth. Composed mainly of oxygen and nitrogen as well as small amounts of other trace gases, the atmosphere is essential to photosynthesis (the process a plant uses light to make food). It also protects organisms from the Sun's infrared rays because it absorbs much of these. Since the atmosphere is located physically between the surface of Earth and the Sun, the molecules of gas that make it up act somewhat like a pane of glass in a greenhouse, which is where the name "greenhouse effect" originates. The glass on a greenhouse lets light in and out but holds in its heat. What the atmosphere does is allow wavelengths of visible light from the Sun to reach Earth's surface. While doing that, however, the atmosphere blocks the escape into space of the longer infrared wavelengths. That is, they trap the light's heat by reabsorbing these wavelengths, much of which get sent back down to Earth again. Overall, this makes Earth a warm place that is hospitable to life.

Human activities have begun to alter this process of capturing heat, however. The result of these activities may be the experiencing of too much of a good thing. Too much of a greenhouse effect means that things are heating up too fast. In the past few decades, human activities have begun to change, if not harm, the atmosphere, so that it is trapping more heat than it should. Specifically, our steady burning of fossil fuels (coal, oil, natural gas) has increased the amount of carbon dioxide in the atmosphere. Carbon dioxide is the most important gas in the greenhouse effect, since it is the molecules of carbon dioxide that do the actual absorbing of long-wave infrared radiation. The more carbon dioxide in the atmosphere, the more heat it keeps in. Besides all of the carbon dioxide being pumped into the air by car exhausts and factories, there is the added problem that hu-

An atmosphere with natural levels of greenhouse gases (left) compared with an atmosphere of increased greenhouse effect (right). (Illustration by Hans & Cassidy. Courtesy of Gale Research.)

SVANT AUGUST ARRHENIUS

Swedish chemist Svant Arrhenius (1859-1927) is considered to be the founder of physical chemistry, a fairly new field which blends chemistry and physics. He not only contributed to the founding of a new branch of science, but added a new concept, now called the "greenhouse effect," to the study of the life sciences.

Svant Arrhenius was a child prodigy (exceptionally smart) who taught himself to read at three years of age. Born at Vik, Sweden, the youngster is said to have become interested in mathematics while watching his father, a surveyor (a person who determines boundaries), add columns of figures. Naturally brilliant in school, he earned a bachelor's degree from the University of Uppsala at the age of nineteen while studying physics, mathematics, and chemistry. For his doctoral dissertation in 1884, he offered a theory that explained what occurred when electricity passes through a solution at the atomic level. His ideas were so revolutionary, however, that his committee gave him the lowest passing grade possible. Nineteen years later, Arrhenius would receive the Nobel Prize in Chemistry for the same work that had barely earned him a degree.

All his life, Arrhenius was regularly pushing the limits of science, and in 1908 he published a book titled *Worlds in the Making* that marked him as one of the forerunners of molecular biology. Molecular biology is the study

mans are vastly reducing the world's forests, whose trees and plants take in carbon dioxide and give off oxygen as part of photosynthesis.

Since humans are artificially warming Earth this way, many scientists feel that Earth is already beginning to experience the negative consequences. Climatologists (scientists who study Earth's climate or weather) now believe that Earth's long-term climate patterns are changing. Some project as much as a 4 or 5 degree increase by the middle of the twenty-first century. This could have disastrous effects, since such global warming could cause the polar ice caps and the mountain glaciers to melt. This could result in mass flooding, making many islands disappear, and putting coastal cities completely under water. Since it is climate that mostly determines what will grow where, crops also could be seriously affected. A global temperature rise could produce entirely new patterns and extremes of rainfall or drought in certain areas.

Steps are being taken to prevent possible disastrous effects from happening. Efforts to reduce the pumping of greenhouse gases (carbon dioxide, methane, and nitrous oxide) into the atmosphere have begun in cer-

of the complex chemicals, like proteins and nucleic acids, that make up living things. In this work, Arrhenius offered what he called the universality of life, meaning that life was not to be found only on Earth. It was his belief that life on Earth had sprung from "spores" that had been driven though outer space by radiation pressure from other planets. This meant to him that life had been spread throughout the universe and took hold wherever conditions were favorable. There are now many physical reasons why his spore theory is not correct, and it also offers no explanation as to where or how life originated in the first place.

It was in this book, however, that Arrhenius also suggested what we now call the greenhouse effect. There he speculated that carbon dioxide gas in the atmosphere heats Earth by first allowing sunlight to reach its surface, and then trapping much of the heat radiation or preventing the heat from escaping back into space. He argued that because of this phenomenon, any rise in the amount of carbon dioxide in the atmosphere would raise Earth's temperature since it would act like a greenhouse and trap even more of the heat. He also argued that the reverse could happen and that a major decrease in carbon dioxide could result in a cooling effect that might even cause another Ice Age. Today, many scientists think that global warming is in fact occurring, and that it is caused by the carbon dioxide released when fossil fuels, like coal, are burned in factories and power plants. Arrhenius was certainly a man ahead of his time.

tain countries. This involves making automobile engines more efficient and clean, or better still, encouraging the use of public transportation. Electric cars soon may become practical. The use of nitrogen-based fertilizers can be reduced, and the destruction of entire forests can be stopped. International agreements have already been made to reduce the production of such gases by industry. Many scientists point to the nearby planet Venus as an example of what could happen if a "runaway" greenhouse effect ever occurs. The atmosphere around Venus keeps its surface temperatures as high as 932°F (504°C). To date, there is no evidence of life on Venus.

[*See also* **Carbon Dioxide; Forests; Ozone; Pollution; Rain Forests**]

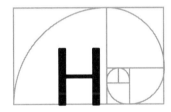

Habitat

A habitat designates the distinct, local environment where a particular species lives. Habitat has also been described simply as a place where a certain organism usually can be found. A shark's habitat is the ocean, and a trout's is a freshwater stream. Habitat requirements of the world's organisms vary widely and can be extremely different even for similar species. Habitat changes can affect the survival of a species.

The habitat where an organism lives is one in which that organism's particular requirements are met. Habitat has been described as the place where an organism can best do its job, or its biological "street address." A favorable habitat for any animal or plant means that this organism is able to get all the things is needs to live, and that it has a slight advantage over others. Habitats are usually described in terms of an outstanding geographical feature such as being mountainous, being a desert, or a being a grassland. However, the concept of habitat is dominated by the idea of a particular type of place. The place where an organism thrives can be anywhere, from the human gut where the bacterium *Escherichia coli* lives, to the soil an earthworm digs, to the rocky crevices a mountain goat leaps across. Habitats can be nearly any size, although when a habitat becomes especially large it is called a biome. The word "environment" is sometimes used in place of habitat, but environment usually suggests a larger area, while habitat suggests a more local area.

Within its habitat, every species eventually finds its niche or the particular role that fits it best. An organism's ecological niche is defined by many factors such as the food it lives on, who are its predators, what temperatures it tolerates, or how much water it needs to consume. While two

THOMAS EUGENE LOVEJOY

A leader in the emerging discipline of conservation biology, Thomas Love-joy (1941–) conducted a unique twenty-year experiment to determine the best-sized habitat for conserving biological diversity (a broad term that includes all forms of life and the ecological systems in which they live). The goal of international conservation also has been fostered by his idea of a "debt-for-nature" swap in which a country's financial debt is partially forgiven if it carries out certain conservation measures.

Thomas Lovejoy was born in New York City into a life of wealth and privilege. While attending the Millbrook School in New York, Lovejoy was first made aware of the wonders of the natural world by that school's founder, Frank Trevor. The young Lovejoy was inspired to study field biology, especially birds, and became fascinated with biology and the natural world. After entering Yale University, he was able to study under the eminent English ecologist, George Evelyn Hutchinson. Lovejoy received his bachelor's degree in 1964. While in pursuit of his doctorate at Yale, Lovejoy was able to spend two years in Brazil as part of a Smithsonian Institution project. There he introduced the technique of bird-banding (tying bits of colored string on birds' legs in order to observe their behavior) to that country, and his doctoral thesis proved to be the first major long-term study of birds in the Amazon region. He was also able to study birds in Africa. In 1971 he received his Ph.D. from Yale University.

After serving the World Wildlife Fund for fourteen years, eventually becoming its vice president, Lovejoy joined the Smithsonian Institution in 1987. There he has been able to continue a long-term experiment that he had begun at the World Wildlife Fund. In 1978, Lovejoy initiated a twenty-year experiment (which has since been extended) to try and determine the best strategy for conserving biological diversity. Called the Minimum Critical Size of Ecosystems (MCSE) Project, its goal is to discover whether bi-

species may live in the same habitat, they can never share the same niche. A niche is a smaller concept than habitat, since a niche is part of the larger habitat. That does not mean, however, that a niche is any less complex. An organism's niche is a highly complex set of activities, impacts, relationships, and roles that can be difficult to understand fully.

Plant and animal habitats are often negatively impacted by humans. It is difficult for human populations to live near a species' habitat and not have an impact on it. Human impact can include climate change, acidification (from acid rain), reduction of wooded areas, lowering of water ta-

ological diversity is better favored by conserving a single large piece of land or by preserving a large number of smaller habitats. Although this was a serious issue that needed to be resolved scientifically, it seemed to be impossible to carry out until Lovejoy came up with an idea. Aware that Brazil had begun a program of developing the Amazon basin, Lovejoy also knew that that country allowed developers to clear only half of the rain forest and leave the remaining half untouched. Seeing that he had the raw materials for a large-scale experiment at hand, Lovejoy was able to get Brazil's support for his new idea. That idea consisted of twenty-four separate reserves, or habitats, that varied in size from 2.5 to 25,000 acres. After two decades of study, the project has yielded valuable results. One of the most interesting is called the "edge effect." This phenomenon occurs when the trees at the edge of the preserved habitat die off and the butterfly populations decline. This suggests that a reserve should always be made larger than is required to support its biological diversity, since there will always be an inevitable shrinkage of life-supporting land around the edges of the preserved habitat. Recent results further suggest that there may also be a loss of diversity deeper into the habitat than first thought.

Lovejoy is one of the first of a new wave of biologists who must combine their science with real-world concerns, such as politics, in order to advance their goals. If a scientist is studying biological diversity and seeks to preserve it, the realities of today's world are such that he or she can no longer remain in the laboratory or jungle and conduct research. Instead, if scientists want to preserve the subject of their study, like the world's essential biological diversity, they are being forced to become actively involved in the politics and economics of the twenty-first century. Lovejoy is a good example of a scientist who has become involved in politics and economics. He has done this through his "debt-for-nature" idea. Today, Lovejoy also is recognized as one of the most effective spokespersons on such issues as global warming and the loss of biological diversity.

bles, and desertification (or the spread of a desert area). Such activities degrade, break up, and eventually destroy entire habitats. These human activities threaten the existence of other organisms. This can be particularly true for species that are highly specialized. An example is the giant panda that feeds only on a certain species of bamboo. Given the rate at which its habitat containing the bamboos species is being destroyed by human development, ecologists do not expect the giant panda to survive in the wild. As a result, giant pandas will eventually only be able to be found in zoos.

[*See also* **Niche; Species**]

Hearing

Hearing is the sense that enables an organism to detect sound waves. It serves mainly to allow an animal to detect danger, to locate its prey, to communicate, and even to express emotion. As one of the five human senses, the ability to hear is an especially important sense because of its connection to human speech and language.

To understand what hearing is about, it is important to understand the nature of sound. Sound needs air to be heard. This means that in a vacuum there is no sound because no air is present. This was proven in the seventeenth century when the English chemist Robert Boyle (1627–1691) placed a bell inside an airtight jar and gradually withdrew all the air as he rang the bell. When all the air was gone, the ringing bell made no sound at all. This is because sound is created by waves of pressure or vibrations that happen when something disturbs the air. It is much like how a pebble thrown into still water creates an increasing ring of ripples. When a noise is made, it disturbs the air and a sound wave begins as the air vibrates from the original noise. Without air to travel through, sound does not exist. A drawing of a sound wave would therefore look like a wavy line.

Only vertebrates (animals with a backbone) and some insects have the ability to hear. This means that only these types of animals have special organs to receive and then interpret these vibrations of the air. Just as smell and taste use chemoreceptors (a nerve cell that responds to chemical stimuli) to detect dissolved or airborne chemicals, and touch and sight use tactile (touch) and visual receptors, hearing employs auditory receptors. Each sense has a specialized type of receptor that is geared to respond to a certain type of stimulus. Most vertebrates have a system that enables them to detect sound waves and then to convert them into nerve impulses that the brain identifies.

HEARING IN MAMMALS

Hearing reached its highest level of development in mammals, a class that includes humans. The human ear is a complicated organ that serves as a good model of how animals hear. Humans and most other mammals have an outer, middle, and inner ear.

The Outer Ear. The outer, or external ear, is called the pinna and is the part that is outside of the head. Its shape is designed to catch and direct sound inwards. Mammals have two ears so they can better locate the direction of a sound. The brain actually calculates the location by compar-

Opposite: A diagram of the hearing process. The out ear collects external sounds and funnels them through the auditory system to the eardrum (far right of illustration). (Illustration by Hans & Cassidy. Courtesy of Gale Research.)

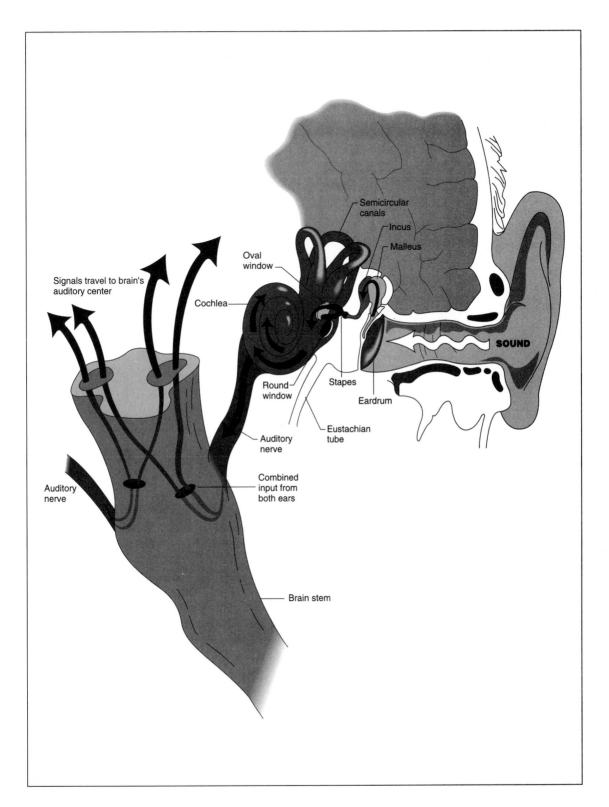

Signals travel to brain's auditory center

Cochlea

Oval window

Semicircular canals

Incus

Malleus

SOUND

Round window

Stapes

Eardrum

Auditory nerve

Eustachian tube

Combined input from both ears

Auditory nerve

Brain stem

ing the speed at which the sound reaches each ear, since it will reach the ear closer to the sound first. As the outer ear collects the sound, it funnels it into a passageway called the auditory canal and then on through to the tympanic membrane or eardrum. This delicate, tight piece of tissue is like the tight skin on a drum. When the sound, made up of certain vibrations, reaches the eardrum, it causes it to vibrate.

The Middle Ear. Beyond the eardrum is the middle ear, and here the vibrations are transferred to three bones inside of it called the hammer, the anvil, and the stirrup. The eardrum makes the hammer vibrate. Like a chain reaction, the hammer makes the anvil vibrate. This makes the stirrup do the same, each time increasing the intensity of the sound.

The Inner Ear. The stirrup then vibrates against the inner ear's oval window, which covers a snail-shaped structure called the cochlea. The cochlea is filled with fluid, which helps the body keep its balance, and is covered with tiny hair cells. It is here that the vibrations somehow get changed into nerve impulses. Once the mechanical vibrations are converted into electrical impulses, they travel through the auditory nerve to the brain's cerebral cortex. There they are interpreted as sounds. The type of sound sensed by the brain depends on which hair cells are triggered.

The Eustachian Tube. It is important for the pressure on both sides of the eardrum to be equal, so the ear has mechanism to do this. Called the Eustachian tube, this tube connects the middle ear with the throat. It is not a permanently open tube but works like a valve that opens and closes as necessary. We experience this pressure imbalance when our ears feel uncomfortable on an airplane and our hearing seems to fade. We also experience the "pop" as we yawn or swallow and suddenly equalize the pressure.

The Cochlea. The ear's cochlea also functions to help people keep their balance as we move about. The fluid inside the cochlea moves and shifts and tells the brain about the body's position. Spinning about makes someone dizzy afterwards because this fluid keeps sloshing about and does not stop all at once. The brain thus receives confusing and chaotic signals until the fluid slows and stops sloshing.

HEARING IN INVERTEBRATES

Most invertebrates (animals without a backbone) do not have specific receptors for detecting these air vibrations that produce sound, and most have no hearing. They do feel the vibrations of the air, water, or soil in which they live. Insects are an exception to the above, since crickets, grasshoppers, katydids, cicadas, butterflies, moths, and flies are capable of hearing. Crickets and katydids have receptors called tympanic membranes, which are similar to ears, on their legs, and grasshoppers and cicadas have them on their abdomen. The tympanic membrane of grasshoppers and crickets are known to function much the way the human eardrum does.

ECHOLOCATION

Some mammals can move their outer ears to better focus on a sound. Certain deep-diving marine mammals like seals and dolphins can close off their ear canals when submerged. Bats and dolphins use echolocation as a way of sensing their surroundings. They send out a sound and then listen for the echo as it bounces off an object. Both can determine not only how far they are from the object, but their brains can actually analyze the echo pattern and form an image of it.

HEARING LOSS

Hearing in humans can be impaired or lost altogether. The most common reason for hearing loss is a stiffening and eventual death of the important hair cells in the inner ear. Loud noises make this happen. It is estimated that the average person has lost more than 40 percent of his or her hair cells by the age of sixty-five. Loudness is measured in decibels, and ears can be permanently damaged by two to three hours of exposure to 90 decibels. Since the music at some rock concerts is often as high as 130 decibels, it is not surprising that some rock musicians and their fans have diminished hearing.

Since communication is such an important part of being human, the loss of hearing is an especially isolating disability for many people. Helen Keller (1880–1968), who became blind and deaf during infancy, said that her lack of sight was nothing compared to how her deafness isolated her from others.

Heart

The heart is a muscular pump that transports blood throughout the body. As an essential part of an organism's respiratory system, the heart circulates blood through the lungs. It exchanges carbon dioxide (gas) for oxygen which it distributes to the rest of the body. The heart is made of tough cardiac muscle that never needs to rest.

All animals that have blood and a circulatory system also need a version of a heart to move that blood throughout the body. The design and complexity of a heart is generally related to the complexity of an organism and to the type of lifestyle it leads. For example, an earthworm's simple system needs nothing more than two pulsating tubes running up and down its body. In vertebrates (animals with a backbone), there is also a degree of complexity depending on the structure of the vertebrate's body. For example, a fish's heart has two separate chambers. One is called an "atrium" that receives blood after it has circulated through the fish's body. The other is called a "ventricle" that forces the blood received from the atrium over its gills. Here the exchange of carbon dioxide and oxygen takes place. Still higher up the vertebrate ladder, the amphibian, like the frog, heart has three chambers, while birds and mammals have four chambers.

THE HUMAN HEART

High-energy animals, like birds and mammals that move about quickly and maintain a constant body temperature, need a very efficient heart. Such a heart can keep oxygen-rich blood entirely separate from blood that has surrendered all its oxygen (deoxygenated blood). The human heart is a good example of how high efficiency is achieved. The human heart is actually two hearts or pumps. One serves the lungs and the other serves the body. Located at the center of the chest cavity behind the breastbone, the heart is about the size of a clenched fist. The surface of the heart is covered with a number of small arteries, known as coronary arteries. These supply the heart's muscle fibers with oxygen-rich blood. The inside of the heart is divided into four chambers. The two at the top, called the atria (singular, atrium), are the receiving chambers. The two at the lower half are called the ventricles and are the pumping chambers. Each atrium is separated from a ventricle by a valve that allows blood to travel only in one direction and prevents any backup. The right and left sides of the heart are separated by a thick wall of muscle called the septum. Blood flows from the right atrium into the right ventricle. The right ventricle pumps the blood to the lungs where it leaves its waste (carbon dioxide) and receives oxygen. From the lungs, blood flows into the left atrium and then to the powerful left ventricle which pumps it throughout the rest of the body.

The heart is made of tireless cardiac muscle that beats, or contracts and relaxes, an average of seventy-two times a minute. This beating is known as the pulse rate. The heart beats in a regular sequence due to a complex electrical network. This network stimulates the muscle fibers and make the chambers contract or pump in the proper sequence. The heart also regulates blood pressure, or the pressure exerted by flowing blood against the inside of the walls of the veins and arteries through which it is flowing.

Opposite: A cutaway view of the anatomy of the human heart. (Illustration by Hans & Cassidy. Courtesy of Gale Research.)

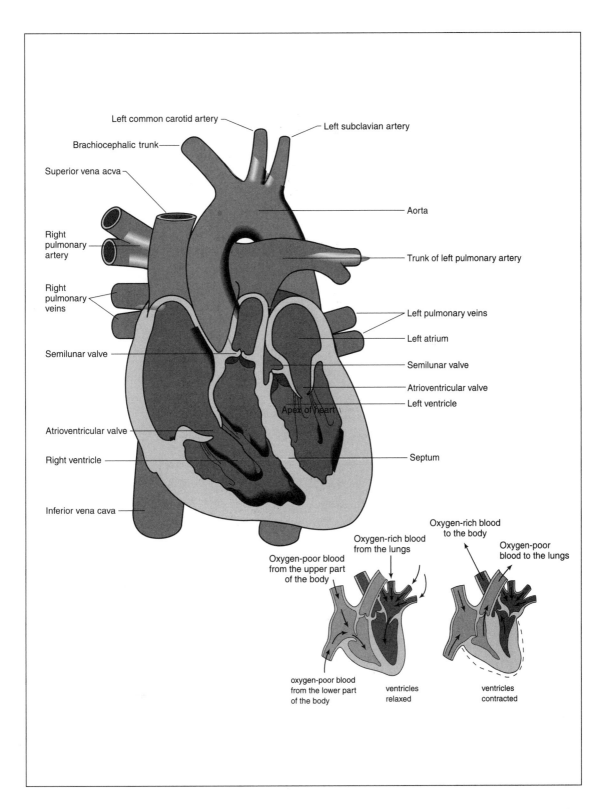

Left common carotid artery

Left subclavian artery

Brachiocephalic trunk

Superior vena acva

Aorta

Right pulmonary artery

Trunk of left pulmonary artery

Right pulmonary veins

Left pulmonary veins

Left atrium

Semilunar valve

Semilunar valve

Atrioventricular valve

Left ventricle

Apex of heart

Atrioventricular valve

Right ventricle

Septum

Inferior vena cava

Oxygen-rich blood to the body

Oxygen-rich blood from the lungs

Oxygen-poor blood to the lungs

Oxygen-poor blood from the upper part of the body

oxygen-poor blood from the lower part of the body

ventricles relaxed

ventricles contracted

The heart is always able to adjust to physical and chemical changes in the body. Its beating strength, rate, and blood pressure are affected by the needs and demands of the body. The heart also responds to chemical changes in the body, such as the increased demands placed upon it during strenuous exercise. In this example, waste products must be eliminated at a much faster rate than when resting. Hormones, the body's chemical messengers, affect the rate at which the heart beats, and the heart is able to react appropriately to all types of stress, whether physical or emotional.

HEART DISEASES AND DISORDERS

Finally, a variety of diseases and conditions can affect the heart. An infection can inflame the sac around the heart (pericarditis), the lining of the heart (endocarditis), or the cardiac muscle itself (pericarditis). The electrical impulses that make the heart beat regularly can be upset. This is called arrhythmia. Also, the arteries that supply the heart muscle can become clogged. This can lead to a heart attack in which a part of the heart muscle dies due to a lack of a proper blood supply. High blood pressure and clogged arteries can also make the heart have to work harder, causing it to enlarge. This condition is known as congestive heart failure, and if not treated, is fatal.

[*See also* **Circulatory System**]

Herbivore

Herbivores are animals that eat only plants. Since animals cannot make their own food, they must obtain their energy either by being herbivorous (eating plants), carnivorous (eating other animals), or omnivorous (eating both plants and animals). Herbivores are called primary consumers on the food chain since they are the first organisms to consume the energy stored by primary producers (plants). As plant eaters, herbivores have certain physical characteristics that make them different from carnivores and omnivores.

Since green plants are the only organisms capable of producing their own food, they are at the beginning of the world's food chain. This food chain or food path connects species in terms of how food and energy is passed from one species to another. Food chains or webs are divided into organisms that produce energy and those that consume it. Producers, who use energy and make their own food, are called "autotrophs." Plants are autotrophs since they make their own food by converting the Sun's light energy into chemical energy via photosynthesis. Animals cannot do this and must eat other living organisms in order to survive. They are called "heterotrophs" since they are forced to obtain their energy and nutrients from the food they ingest or eat. Animals are considered to be primary, secondary, or tertiary consumers according to the order of who eats whom. A caterpillar is a primary consumer because it eats only plants. A pigeon would be a secondary consumer when it eats a caterpillar. A fox would be a tertiary consumer when it eats a pigeon.

Herbivores have developed many traits that are specialized only for plant eaters. Cows and sheep are examples of herbivores that possess several physical characteristics because of their diet. An examination of their skulls shows that they have no canine teeth, or the two long, pointy teeth in the front of a carnivore's mouth that it uses for ripping and holding its prey. Since these herbivores are grazing animals and have no need to hunt and catch their food, they have instead chisel-like front teeth called incisors to break off blades of grass. Behind the incisors and the back molars is a gap or a space called a diastema, providing the necessary "give" in a jaw that moves side-to-side. These herbivores' molars are flat teeth with ridges on the surface that serve as powerful grinding tools. Sheep and cows also have loose joints in their jaws so they can chew side-to-side, which improves the grinding action of their teeth. After herbivores chew the leaves or grass into a pulp with their molars, it passes into a highly specialized digestive system that is very different from that of a carnivore. The best example is that of a cow, since it has four chambers in its stomach. Grass is very difficult to digest because of its tough cell walls. Sometime after the pulp enters a cow's first chamber or rumen, it is regurgitated or coughed up into its mouth to be chewed again. This helps break the grass down even more and is what we describe as a cow "chewing its cud." The grass later passes on through the remaining chambers of a cow's stomach where it is digested even more. Since herbivores are able to get only a small amount of energy from each mouthful of their vegetarian diet, they must eat enormous amounts of food. This is why a cow spends nearly all its waking hours grazing.

Besides cows and sheep, other herbivores include caterpillars, fishes, birds, and many other mammals. Many eat seeds and fruit instead of grass or leaves, and because of this specialized diet, have specialized tools, like a certain type of beak or bill. Certain finches have beaks that allow them to eat a certain type of seed. Other birds like the toucan have long, sharp bills to pluck berries or chop larger fruit into bite-size pieces. Herbivores are critically important to carnivores, since without them, a meat-eater would have no way of obtaining the life-giving energy first captured by green plants.

Herpetology

Herpetology is the scientific study of amphibians and reptiles. As one of the many subfields of vertebrate (animals with a backbone) biology, it focuses on the anatomy (structure), physiology (processes), behavior, genetics, and ecology of amphibians and reptiles. Although amphibians are very different from reptiles, they are grouped together in the discipline called herpetology.

Amphibians live both on land and in the water, and need to keep both their skin and their eggs moist. Frogs, toads, newts, and salamanders are amphibians. Reptiles are best suited to life on land, and both their skin and their eggs have tough outer coverings. Turtles, snakes, lizards, and crocodiles are reptiles. Both amphibians and reptiles are ectothermic (cold-blooded), meaning that their temperature matches their environment. The grouping together of these different types of vertebrate animals into one field is believed to have come from the very old tradition of lumping together all creeping or crawling organisms. In Greek, *herpeton* means a crawling thing and *logos* means reasoning or knowledge. The first book to systematically arrange animals in some sort of order was written by the English naturalist, John Ray (1628–1705), in 1693, and in that work he grouped amphibians and reptiles together.

English naturalist John Ray wrote the first book that systematically arranged animals. In this work, however, Ray incorrectly grouped amphibians and reptiles together. (Photograph courtesy of The Library of Congress.)

A herpetologist, or one who practices herpetology, must be familiar with a wide variety of animals. Amphibians (whose name in Greek means "having two lives") undergo a unique process called metamorphosis. This is a dramatic but natural change in body shape that transforms an organism (like a tadpole that lives totally underwater) into an entirely different organism (like a frog that breathes air). Reptiles are often confused with amphibians from which they evolved, but they do not undergo metamorphosis and breathe air through lungs their entire lives. Amphibians must return to water to reproduce, and fertilization occurs externally (outside of their bodies). Reptiles produce a sealed egg that hatches on land, and the eggs are fertilized internally, or within the body of the female. All amphibians and reptiles belong to the phylum Chordata and the subphylum Vertebrata. Both also belong to a superclass called Tetrapoda. A superclass is a name devised by taxonomists (biologists who specialize in classifying things) when an extra classification group is needed. It is similar to the categories called superfamily and subphylum, which are also used when the realities of life do not completely fit into the standard seven-part classification scheme.

The study of herpetology has a significance beyond the animals themselves and extends to humans as well. Since amphibians and reptiles are easily kept in captivity and have systems that manage and respond to foreign and toxic substances in much the same way that human's do, they are ideal subjects for studies that benefit people. In fact, because of this similarity, many scientists believe that the health of both (especially amphibians) may serve as an early warning system for the overall health of our environment. Thus, if frogs become unexplainably sick, it may be a warning that there is something harmful in the environment, which may eventually cause health problems in humans.

[*See also* **Amphibians; Reptiles**]

Hibernation

Hibernation is a special type of deep sleep that enables an animal to survive the extreme winter cold. Hibernation lowers an animal's energy needs and allows it to live off stored fat and not have to search for scarce food. Hibernation is a form of cyclic behavior and is triggered by different cues in the animal's environment.

All animals have different survival tactics that allow them to live through difficult or life-threatening situations. The steady, severe cold that comes with winter poses a problem to animals who do not escape it (by migrating or leaving) nor adapt to it (such as by growing an extra-thick coat of fur or a layer of fat). Winter is difficult for all warm-blooded animals (those that maintain a constant internal temperature despite their environment), since they must spend most of their energy just keeping warm. When the temperature falls below freezing, these animals must eat even more than they usually do simply to produce enough internal heat to stay alive. Nature makes things even more difficult in winter, since at a time when warm-blooded animals need to increase their intake of food, it has suddenly become very scarce.

For certain animals, hibernation is a simple way to solve the particular problems posed by winter cold, since they basically sleep through winter and wake when the weather has become mild. However, hibernation is a fairly complicated physiological event. When an animal hibernates, its body processes such as breathing and heartbeat slow down, sometimes to the point that the animal appears dead. While hibernation could be described simply as a very deep sleep, it is anything but simple. True hibernators are usually small mammals like woodchucks, mice, or ground squirrels. When the time comes to hibernate, the animal responds to one of several environmental "cues," such as a certain low temperature or a reduction in the hours of daylight. These cues trigger the release of a hormone (a chemical messenger) called hibernation induction trigger (HIT) that causes major changes in the animal's body. Its heartbeat becomes slow and weak and its body temperature drops many degrees. It takes a few breaths every minute and it makes hardly any waste (urine). As a true hibernator, the animal falls into such a deep sleep that it looks dead and sometimes cannot even be awakened if picked up. All of these reactions triggered by the hormone allow the animal to maintain its necessary body processes while using far less energy than if it were awake.

Before they fall asleep for the season, hibernators usually develop huge appetites that allow them to store as much fat as possible to be burned later while sleeping. They also usually prepare the den or burrow where they sleep to comfortably insulate themselves from the cold. As spring approaches, different cues in their environment, like warmer temperatures or lengthening daylight, awaken them and they soon resume their normal level of activity. This does not happen immediately, however. As their heart rate increases, along with their blood pressure and respiration, hibernators usually begin to shiver, which slowly raises their body temperature. After readjusting to this now-high rate of metabolism (the chemical processes that take place in an organism), they are ready for normal activity.

Like their warm-blooded counterparts, cold-blooded animals (whose temperature changes with the surroundings) also hibernate. Animals like frogs, turtles, and snakes bury themselves in the mud where their sloweddown systems find just enough trapped oxygen to stay alive. Some insects like butterflies also hibernate in the open, and their systems produce chemicals that act as antifreeze.

Other animals like bears, raccoons, and skunks are not true hibernators, although they do very much the same thing. Rather, they are considered "light sleepers" since their bodily functions do not fall to such low rates as those of true hibernators. These "light sleepers" also sometimes wake up and eat something that they may have stored or even go outside to eliminate waste. Brown bears can awaken very quickly, and often give birth during a long, cold winter. Even true hibernators have built-in mechanisms to awaken them under unusual conditions. For example, if temperatures fall to such a low that even the hibernating animal is in danger of freezing to death, its body will automatically switch to actively producing heat.

Another form of hibernation occurs in the summer. This is called "estivation," and is a process similar to hibernation. Certain desert animals estivate underground when they are threatened by prolonged, extreme heat or drought. Estivation is what enables many desert dwellers to survive during the very hot summer months.

[*See also* **Metabolism**]

Homeostasis

Homeostasis is the maintenance of stable internal conditions in a living thing. Organisms use a variety of systems and processes that help regulate and maintain a constant environment within their bodies. All organisms use a self-adjusting balance to make sure that what is going on inside their bodies is kept within certain boundaries.

One of the main characteristics of living things, or organisms, is that they have the ability to adjust to their environment. In the highly competitive struggle for survival, an organism would be at a great disadvantage if it could not adjust and be ready to cope with a changed situation. Such change can occur inside or outside an organism. Since living things are extremely complex organisms with constant energy demands, there are countless cellular reactions going on all the time. For example, chemicals are combining and breaking apart, fluids are passing in and out of membranes, and substances are being converted from one form to another. All of this constant activity means that the environment inside an organism's body is dynamic, and that it is always in a state of movement and change.

CLAUDE BERNARD DEVELOPS THE CONCEPT
OF HOMEOSTASIS

The concept of homeostasis was developed by the French physiologist (a person specializing in the study of life processes, activities, and functions) Claude Bernard (1813–1878). Bernard investigated how the body keeps itself in a stable, or steady state. It was Bernard who first recognized the idea behind homeostasis—that an organism is designed and operates on the principle that it will always attempt to maintain a balance in its systems.

After Bernard, science eventually discovered that living things use two simple self-adjusting elements, input and output, as regulators. Although these elements are uncomplicated, an organism has many mechanisms and structures that it uses to maintain homeostasis. Some of these mechanisms work automatically and are under the control of the autonomic nervous system. It is this system that regulates the body's involuntary processes like internal body temperature, blood pressure, and food digestion, among other function.

Certain changes in the external environment may automatically trigger an organ to take a certain action. Under certain conditions, our bodies will perspire whether we want them to or not. Other mechanisms are controlled by the body's endocrine system. This system uses chemical messages, known as hormones, inside the body to regulate functions. A hormone is secreted by the organ that produces it when something happens to the organism that warrants regulation. Hormones then travel through the bloodstream to their target cells, which take the appropriate action.

THE FEEDBACK SYSTEM

For this system to really work, however, it must have some way of getting updated information as to what is going on inside and outside the organism. This is achieved by a feedback system that operates something like the thermostat in a house. Every thermostat has a sensor that, when set at a certain temperature, automatically turns the furnace on when the temperature gets lower than its setting, and turns it off when it reaches it. In this way, a built-in feedback "loop" regularly monitors its environment and is able to maintain a constant temperature by turning the furnace on or off.

The feedback system in our bodies works mostly with what is called negative feedback. Negative feedback operates by detecting an unwanted change and countering it in order to balance it. A typical example would be the body's use of feedback mechanisms to raise or lower its internal temperature during extremely cold or hot weather.

LEVELS OF HOMEOSTASIS

A diagram showing a negative feedback loop regulating blood pressure. (Illustration by Hans & Cassidy. Courtesy of Gale Research.)

In the human body, homeostasis takes place at many different levels. These include the molecular level (in which atoms are linked together by chemical bonds), the cellular level, the organism level, and the population level. An example of homeostasis at the molecular level would be the body limiting how much of something is produced by a certain chemical reaction. At the cellular level, an example would be how certain cells

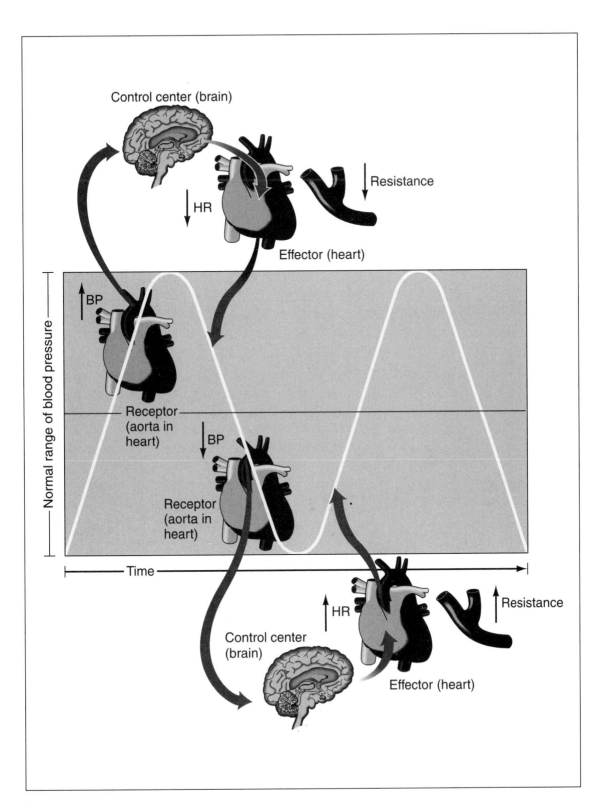

Control center (brain)

Resistance

↓HR

Effector (heart)

↑BP

Normal range of blood pressure

Receptor (aorta in heart)

↓BP

Receptor (aorta in heart)

Time

↑HR   Resistance

Control center (brain)

Effector (heart)

will stop dividing if they become so numerous that they touch each other. Sensations of hunger and thirst are good examples of a homeostasis mechanism at the organism level. Finally, at the population level, an increase in the number of prey animals usually results in an increase in the number of predators that eat the prey. In this way, the population size of both animals is kept in balance.

[*See also* **Endocrine System; Hormones**]

Hominid

A hominid is a family of primates that includes today's humans and their extinct direct ancestors. Before humans evolved into what they are today, several human-like species existed, some of which went extinct and some of which evolved into today's species—the only living species of hominids. Fossil discoveries suggest that the complete story of human evolution is still not fully known, although many of the major hominids species are documented.

The word hominid, which includes *only* human beings and their direct or immediate ancestors, should not be confused with the similar word hominoid. Hominoid includes *both* humans and apes, and therefore refers to a much larger and more diverse group of primates. All hominids are hominoids, but not all hominoids are hominids.

The first hominids are thought to have appeared on Earth about 3,000,000 to 4,000,000 years ago. The earliest known fossils have been discovered in southern and eastern Africa, and all appear to have three features in common: bipedalism or upright walking; an omnivorous diet (plant and animal); and an expansion of the brain. Eventually, over very long periods of time, these and other biological changes occurred and hominids became less and less apelike and more like today's modern humans. Many think that the first or earliest hominids came down from living in the trees and moved into the open fields or plains. Some think that a major climate shift brought this about since it is known that when hominids first appeared, the savannas were starting to replace forests. This is believed to have forced hominids to make a transition from being forest and tree dwellers to living in a more mixed habitat with woodland and open grasslands.

It is important to take note of the actual physical changes that would make hominids different from other primates. One obvious difference were different facial features. Hominids would lose much of their muzzle or protruding jaw, and the overall size of their faces would be reduced, especially their teeth and jaw. Their teeth would also develop thicker enamel and become less specialized. This is an indication that their diet was also less specialized and was probably omnivorous (both plant and meat eaters). Also reduced would be the bony ridges over their eyes, and the back of the skull would lose its crest or raised edge. The brain would also become larger in comparison to the rest of the body. At some point in their transition from trees to plains, hominids became bipedal, meaning that they could walk upright on two feet. This not only made them taller and able to see farther, but left their arms and hands free to carry things, to use tools, or otherwise do things that would further promote their survival. Bipedal walking resulted in significant changes to a hominid's lower spine, leg bones, and pelvis.

The oldest known hominid was found in South Africa and is called *Australopithecus ramidus.* Dated at about 4,400,000 years ago, this species walked on two legs but had a fairly small brain. There probably were even earlier types of hominids, but no one has yet found fossil remains. Paleoanthropologists (scientists who study the fossil remains of hominids) are not sure if *Australopithecus* is our direct ancestor or not. The first hominid to be considered human and therefore given the genus name *Homo* appeared probably about 2,000,000 years ago. Called *Homo habilis* meaning "handy man," it had a much larger brain than *Australopithecus* and is known to have used stone tools. Its skull and teeth were also different, and its face was smaller and more in proportion with the rest of its body. Between 1,500,000 and 500,000 years ago, *Homo habilis* was either replaced by or evolved into *Homo erectus* or "upright man." This is believed to be the first hominid to venture out of Africa and move into Asia and Europe. Significantly, its brain was even larger and it was able to use fire and make hand axes. About 300,000 years ago, the first *Homo sapiens neanderthalensis,* or Neanderthal man, appeared. Although it had a brain as large as humans are today, its head was still different, as its eye ridges were heavy, probably making it look fierce. Neanderthal man also made tools but unlike *Homo habilis* buried its dead in special graves.

About 40,000 years ago, humans similar to today's species first appeared. Called *Homo sapiens sapiens* ("wise man"), they may have inter-

British paleoanthropologist Louis Leakey holding skull fragments of an early hominid he discovered in Africa. (Photograph courtesy of The Library of Congress.)

LOUIS SEYMOUR BAZETT LEAKEY

British paleoanthropologist Louis Leakey (1903-1972) was a pioneer in the field of paleoanthropology, which is the study of the fossils of early humans and prehumans. He discovered the earliest known hominid (a family of primates that includes humans and the immediate ancestors of humans) and showed that humans were not only older than previously believed but that they may have first evolved in Africa.

Louis Leakey's parents were missionaries who were trying to convert African natives to Christianity. Leakey was, therefore, born in Kabete, Kenya, which was then part of the British Empire. He was raised among the Kikuyu tribe, a group of Africans who lived in the area where the mission was located. The young Leakey was able to speak the Kikuyu language as well as his own English, and although he had a governess who instructed him, he spent most of his time with other Kikuyu children exploring the countryside. This would remain with him all his life, and it is said that Leakey always thought of himself as an African instead of an Englishman. When Leakey was finally sent to England at age sixteen to begin his formal education, he found he could not get along with the typical English schoolboy with whom he had nothing in common. Although he got along better at Cambridge University, he was forced to take a year out of school when he suffered a head injury when kicked twice in a rugby match. This absence from school enabled him to join a fossil-hunting expedition to Tanganyika (now Tanzania), an experience that showed him what he really wanted to do in life.

After Leakey obtained his degree from Cambridge in 1926, he decided to devote his career to studying the origins of humanity, which he believed would be found in Africa. At this time, most scientists believed that Asia and not Africa was the original center of human evolution (the process by which humans changed over generations). Leakey began his work at two African fossil sites, one at Lake Victoria and the other at Olduvai Gorge, now in Tanzania. Olduvai was a 350-mile (217.36 kilometers) ravine that contained a great deal of evidence, like primitive stone tools, that some forms of humans had lived there very long ago.

bred with Neanderthals, or Neanderthals may have simply died out. This newest species began using its brain in ways not seen before, made better tools, began cultivating crops, and created sculptures and cave paintings. They developed language, music, built cities, and eventually created civilizations. All of these and other activities are suggested when we say that humans developed culture. Since today's particular species burst on

During the mid-1930s, Leakey divorced his wife and married one of his students, Mary Douglas Nicol (1913-1996). Together, they would spend more than thirty years at Olduvai searching for the fossil remains of the creatures who had made and used those tools. The Leakeys were very determined scientists and put up with a great deal of hardships at Olduvai. They seldom had enough financial support and the remoteness of the site made their supplies and equipment scarce and difficult to haul. Finally, in 1959 while Leakey himself was in his tent sick with malaria, Mary discovered the fossil they had been looking for. She located the skull fragments of a hominid with a small brain and near-human teeth that they named *Zinjanthropus boisei* and later renamed *Australopithecus boisei.* This was the first more or less complete skull of its kind, and it was also the first to be accurately dated. Potassium-argon testing showed that it was about 1,800,000 years old. Although Leakey argued it was probably an evolutionary deadend and not a direct ancestor of modern humans, it nonetheless added considerably to the knowledge of human origins and showed that humans are older than previously thought.

The following year, Leakey's son, Jonathan, discovered the fossil remains of the larger-brained *Homo habilis,* or "handy man," which Leakey claimed was the direct ancestor of modern *Homo.* For this claim, Leakey received a great deal of criticism, and it must be said that he often would overstate his claims and overpublicize himself and his work. Leakey was an ambitious man who recognized the value of publicity in terms of obtaining financial support for his work. Despite his sometimes overblown claims, his significance resides in the fact that he did change the views concerning human development and pushed back the date when humans first appeared to a time much earlier than scientists had originally thought. He also showed that human evolution began in Africa rather than in Asia, which was also an early belief. As recently as 1977, five years after his death, his wife Mary discovered a set of footprints that were dated to about 4,000,000 years ago. After Mary died in 1996, the Leakey's son, Richard, continued their work.

the scene some 40,000 years ago, too little time has passed for us to notice any real biological changes, and any evolution that humans have made since then has been primarily cultural rather than biological.

[*See also* **Fossils;** *Homo sapiens neanderthalensis;* **Human Evolution**]

Homo sapiens neanderthalensis

Homo sapiens neanderthalensis, commonly referred to as Neanderthal man, is a species of the hominid (human) family *Homo sapiens* that disappeared about 30,000 years ago. As an early member of this species, Neanderthals were shorter and stockier than today's humans and had differently shaped heads with heavy ridges over the eyes, although their brains were as large as that of modern humans. Neanderthals existed about the same time as modern man emerged, and it is not known whether they were assimilated into the new group by interbreeding or were somehow made extinct by violence or disease.

An illustration of an early Neanderthal man. Scientists still are not sure whether Neanderthals are ancestors to modern humans or not. (Reproduced by Corbis Corporation (New York).)

The first fossil finds of Neanderthals were made in Germany in 1856, and simply by studying the heavy-ridged brows of the skulls, it was realized that if these bones were human, they were those of a distant ancestor. Eventually these and other bones were dated to between 70,000 to 35,000 years ago, and their rugged bodies indicated that they had adapted to an existence in a cold climate. For some time, Neanderthals were considered to be a form of prehuman brutes, but the size of their brains was shown to be as big or bigger than modern man's. When Neanderthal stone tools and weapons were later found that were more advanced than their predecessors, *Homo erectus,* it was realized that Neanderthals were not as primitive as believed. Neanderthals also demonstrated the beginnings of certain cultural activities that would become a human trademark. One of these was the simple fact that they buried their dead in special graves, suggesting that they had some awareness and sensitivity to the permanent loss of an individual.

Neanderthal fossils have also been found in Asia dating from as far back as 125,000 years ago to as recent as 35,000 years, and aside from anatomical differences in their pelvis, shoulder blades, and skull, they did not look that much different from modern humans. Their skulls, however, did show heavy eyebrow ridges and facial bones. This would have made their facial features much less delicate or refined than those of today's humans. The contents of their skulls, however, were similar, and their brain size ranged between 1,300 and 1,750 cubic centimeters (512.2 to 689.5 cubic inches), much like modern man's. Neanderthals had sophisticated tools and they probably lived in caves and rock shelters. Today, there is disagreement as to whether Neanderthals are part of the gene pool that gave rise to modern humans. A controversial comparison of Neanderthal DNA obtained from fossil bones has not proved conclusively whether Neanderthals were our ancestors or whether they were a dead-end branch on the human tree.

Researchers had previously thought that human speech began about 40,000 years ago when *Homo sapiens sapiens* emerged. However, a recent study argues that for thousands of years prior to this, Neanderthals had the ability to speak. This claim is based on the diameter of a nerve canal that connects the brain and the tongue in Neanderthal skulls. This "hypoglossal" canal is roughly the same size as that in a modern human skull, and implies that Neanderthals may have had the necessary physical equipment for speech. Whether they were modern humans' direct ancestors, an evolutionary deadend, or a species that succumbed to disease or slaughter, researchers continue to study Neanderthal fossils in order to educate themselves about human evolution.

[*See also* **Fossil; Hominid; Human Evolution**]

Hormones

Hormones are chemical messengers found in both animals and plants. In animals, hormones are produced by glands and travel through the blood to certain target tissues. There they act as chemical regulators. Hormones influence reproduction, growth, and overall bodily balance, among other things.

Hormones are important to both plants and animals, but especially to animals. Hormones regulate key bodily functions like body growth, sexual maturity, reproduction, and the maintenance of a stable, or balanced, internal environment. Some hormones have a temporary effect, such as those that regulate the body's blood sugar level. Others cause permanent changes, such as those that make a person grow tall and mature sexually. Still others are present only in certain situations, such as those that prepare a body for stressful situations. Whatever their effect, hormones help an organism to adapt to its environment in the best manner possible. The word hormone comes from the Greek *hormaein* meaning to excite or to set into motion, and this describes what hormones do—they have a stimulating effect.

In vertebrate animals (animals with a backbone), hormones are produced by certain glands, tissues, or organs. They travel through the circulatory system (a network that carries blood throughout an animal's body) to target cells. Hormones do not produce an effect until they reach these specifically receptive cells. The target cells are programmed to react when stimulated by a certain hormone. Only the target cells in the target organ are able to produce the desired effect, since they have receptors that recognize and bind to the hormone.

THE ENDOCRINE SYSTEM

It is estimated that vertebrates have at least fifty different hormones, and many are produced by what is called the endocrine system. Some of the major glands in the human endocrine system are the pituitary gland and the pineal gland at the base of the brain; the thyroid and parathyroid in the throat; and the adrenal, gonads, thymus, and pancreas in the trunk or lower half of the body. Each of these endocrine glands releases its own particular hormone into the bloodstream and each produces the desired effect when it reaches the appropriate target cells. Thus, some hormones

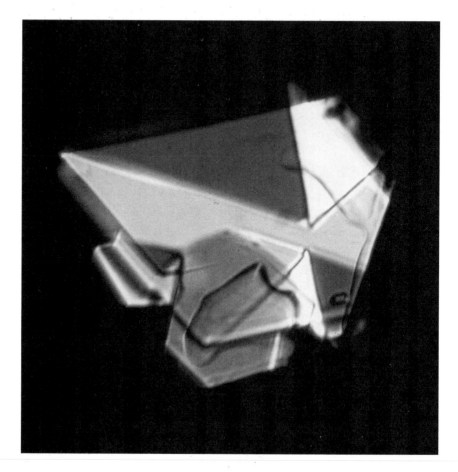

A beef insulin hormone. Most hormones fall into two main categories called peptides and lipids. (Reproduced by permission of Phototake.)

stimulate the growth of muscle and bone, others begin the secretion of milk, and still others influence blood pressure. All are usually produced in tiny quantities, yet all have profound and major effects on the body.

HORMONES IN HUMANS

For example, female sex hormones, like the group called estrogens, are produced by the pituitary gland as well as the ovaries. These hormones produce the dramatic physical changes that take place when a young girl starts to become a young woman. The estrogens trigger the development of what are called secondary sexual characteristics, like breasts. Later, if a woman becomes pregnant, these and other hormones prepare her body to carry and nourish a developing fetus. The male sex hormone, testosterone, is made by the testes and produces the typical male secondary sex characteristics during puberty.

Another well-known hormone is adrenaline. This produces what is known as the "fight-or-flight" response. When an individual senses danger, the body automatically enters a state of readiness to either fight for survival or to take measures to avoid a conflict. This powerful hormone works with great suddenness and a person can actually feel its effects. The heart rate and blood pressure quicken, the skin goes pale, the body's blood sugar level rises, and a person's strength increases. All of these and other physiological reactions take place immediately and without the person's conscious will. These reactions give people a better chance to act immediately and possibly survive a threat.

In humans and animals, hormones are made of either proteins or steroids (which are a type of lipid or organic compound that includes fats, oils, and waxes). The adrenal gland and the gonads (male testes and female ovaries) produce steroid hormones, while the rest of the endocrine system makes hormones that are protein-based and, therefore, are made out of amino acids (the building blocks of proteins). Abnormally functioning endocrine glands can result in too much or too little hormones. For example, a lack of human growth hormone from the pituitary gland can result in dwarfism, while too much can produce giants who suffer from a condition known as acromegaly.

HORMONES IN OTHER ORGANISMS

While other living things also have hormones, their hormones are nowhere near as dominating an influence as they are for animals. Other organisms do not have as elaborate a system of transport and reception as do vertebrates. However, hormones are still very important to invertebrates (animals without a backbone). Invertebrates use hormones mainly

ERNEST HENRY STARLING

English physiologist Ernest Starling (1866–1927) helped create endocrinology, a major branch of medicine and physiology that studies the glands of the body. Starling not only discovered the digestive hormone secretin, but he also suggested the name "hormone" to describe the body's chemical messengers.

Ernest Starling was born in London, England. His father worked in Bombay, India, for the British government and was able to come home only once every three years. The young Starling was educated at King's College School in London and then enrolled at Guy's Hospital Medical College in London. While there, he took advantage of the opportunity to do research in Heidelberg, Germany, and study with the eminent German physiologist, Wilhelm Kuhne (1837–1900). After obtaining his medical degree in 1889, he was appointed demonstrator of physiology at Guy's, and he eventually became head of the department of physiology there. By the time he left there in 1899 to become Professor of Physiology at University College, he had made a name for himself studying the conditions that cause fluids to leave blood vessels and enter the tissues. In fact, in 1896 he demonstrated a phenomenon that came to be known as "Starling equilibrium."

Starling is best known, however, for his work with the English physiologist William Maddock Bayliss (1860–1924), who became his brother-in-law in 1893 when he married Starling's sister. Together they began a study of the secretion of digestive juices by the pancreas (a gland). The normal pancreas

in their growth and development. For example, insects that molt, or periodically shed, their skin produce a hormone that allows this to happen at the right time. Metamorphosis (the complete bodily change that takes place in an insect, such as when a caterpillar changes into a butterfly) is controlled by hormones. When an octopus changes its color during stress, it is a hormone that causes this dramatic reaction.

For plants, hormones allow them to react to the changing conditions of their environment. Some hormones promote cell division, others stimulate or slow growth, and others cause a plant's fruit to ripen. Plants do not have specialized structures for hormones as animals do. In fact, a plant can even be affected by the hormone of its neighbor. This sometimes occurs when a plant releases the ripening hormone called ethylene into the atmosphere. Thus, a fruit like an apple continues to produce this ripening hormone even after it is picked, and will therefore speed up the ripening of any other fruit nearby.

releases several different juices into the duodenum (the top part of the small intestine) to assist digestion. After they had cut all the nerves to the pancreas, they found that the organ continued to release its juices. This proved that it was not under nervous control (that is, not controlled directly by the brain), and so they concluded in 1902 that it must have received a chemical rather than an electrical message. This meant that the message must have been sent to the pancreas through the blood when food entered the duodenum.

They soon found that the small intestine secretes a substance, or a chemical messenger, into the blood that they named "secretin." Further research showed that the secretin was released under the influence of stomach acid. This was the first time that a certain chemical had been proven to act as a stimulus for an organ that was located somewhere else in the body. Starling and Bayliss eventually came to call any chemical that transmits a message from one part of the body to another part a "hormone." This word was taken from the Greek root meaning "to excite." Although hormones had actually been known before the discovery of secretin in 1902, it was Starling who first clearly defined the concept in 1905 and who detailed the role that such substances play in the body. It was thought that Starling and Bayliss were strong candidates for a Nobel Prize, but World War I (1914–18) intervened, and no awards were given for those years. As for recognition from his own country, Starling had been such an outspoken critic of the way his country had managed the war effort that he was given no awards in his lifetime.

[*See also* **Endocrine System; Reproduction System**]

Horticulture

Horticulture is a branch of agriculture that deals with fruits, vegetables, and ornamental plants. It includes the production of fruits and vegetables for food, and the use of plants in landscaping and decorations.

The word horticulture comes from the Latin words *hortus* meaning "garden" and *colere* meaning to cultivate, and was first used in England in 1678. The word *hortus* or garden is an important part of the idea of horticulture, since the concept of the garden as being different from the open field dates back to the Middle Ages (500–1450). During this era, there were three types of areas where things grew. First were the large, open fields where farmers raised mostly grain and fiber crops. Next came the garden or *hortus* which meant a much smaller space that was intensively cultivated with plants used mainly in the kitchen. Finally there was the forest where timber and wild game were found. Today, horticulture includes the art and science of gardening, and is closest to the second of these categories. However, modern horticulture has gone beyond the tiny kitchen garden and has become an entire industry. It is from this industry that people obtain the fresh fruits and vegetables that they eat, the flowers they use to beautify our environment, and the trees and shrubs they use to decorate the outside of their buildings. Although the notion of intensive gardening in a fairly small space distinguishes horticulture from agriculture (which is large-scale), the boundary between the two becomes less clear with an activity like commercial vegetable production.

Modern horticulture is usually divided into two large categories: food crops (olericulture and pomology) and ornamentals (floriculture and ornamental horticulture). Olericulture deals with vegetables grown for food, and pomology deals with fruit and nut crops. Floriculture is concerned with the production of flowers and potted plants, while ornamental horticulture deals with the use of trees, bushes, shrubs, and grass in outdoor landscaping. However, no matter which aspect of horticulture is being practiced, the gardener or grower must be familiar with all the factors that may increase or decrease a plant's growth and development. While the growers need not be botanists (people who specialize in the study of plants), a great deal of serious horticultural research goes on at colleges and universities, agricultural experiment stations, and botanical gardens from which growers benefit.

Human Evolution

Human evolution is the process by which the modern species of humans was formed and developed. Although the major stages of human evolution are known, there are large gaps in our knowledge. It is now widely accepted that apes and humans evolved from the same ancestor. Many scientists believe that the earliest modern humans evolved first in Africa and then spread throughout the world.

Today's human beings, or *Homo sapiens sapiens,* belong to the hominid family tree. Hominid means "human types," and describes early creatures that split off from the apes and took to walking upright, or on their hind legs. In the overall history of life on Earth, the human species is a very recent product of evolution (the process by which living things change over generations). Since there are no human-like fossils older than

4,000,000 years, this makes us only one-thousandth the age of life on Earth. Today, most scientists agree that humans and apes evolved from the same ancestor. Where scientists cannot agree is how long ago this happened, where it happened, and how it happened. However, it has been demonstrated that the difference between the human genetic (heredity) code and that of a chimpanzee is only 0.7 percent.

It is not difficult to pinpoint what separates today's humans from the apes (gorillas, chimpanzees, orangutans, and gibbons). Some of these distinctions are more significant than mere physical difference. Humans have bigger brains and are intellectually superior. They have used this brainpower to develop tools, which have helped humans adapt to nearly any environment in which we choose to live. Humans are bipedal, meaning that they are able to walk on two feet, leaving their arms and hands free to make tools and do things. Finally, humans have developed language, a very sophisticated form of communication. But how did they separate from the apes and begin our own distinct evolutionary path toward what humans are today?

FOSSIL EVIDENCE

It is known humans evolved from the same ancestor as the apes. However, genetic evidence points to apes and humans going in different paths on the African continent between 10,000,000 and 6,000,000 years ago. Scientists have been able to assemble an impressive amount of fossil evidence that documents some of the branches on the hominid family tree.

Australopithecus afarensis. The oldest hominid scientists know about is called *Australopithecus afarensis.* It lived about 3,000,000 to 4,000,000 years ago and stood about 3.5 to 4 feet (1.07 to 1.22 meters) tall. It had a brain the size of a chimpanzee but it was clearly built for upright walking.

Homo habilis. Many scientists believe that the next known stage, *Homo habilis* who appeared about 2,000,000 years ago, came from *Australopithecus.* As the first hominid to be given the genus *Homo,* or "man," it was taller and used tools made of stone. Its name therefore means "capable man" or "handy man."

Homo erectus. By 1,500,000 years ago, it is known that a new, taller human species appeared and possessed a brain that was about half the size of humans today. Called *Homo erectus,* or "upright man," it was the first hominid to use fire and hand axes and to substantially move about. In fact, it is thought that this species left Africa and gradually migrated into Asia and parts of Europe.

Homo sapiens. *Homo sapiens,* or "wise man," are thought to have appeared about 400,000 years ago. With a still-larger brain and more human features like smaller teeth and a clearly defined chin, it used more sophisticated tools and buried its dead.

Neanderthals and Cro-Magnons. One form of *Homo sapiens,* known as Neanderthals, eventually died out. Some argue that Neanderthals evolved into the Cro-Magnons. Others say that the Cro-Magnons were a separate species and simply out-competed or killed the Neanderthals. In 1997, genetic studies of a Neanderthal bone revealed no evidence of a genetic connection to modern man, suggesting that they may have been a separate and distinct species that went extinct. Nonetheless, it was the Cro-Magnons who survived and continued to reproduce. They were humans who mostly looked like us, although with possibly broader faces, and they had moved into Europe. Their population increased rapidly and they began to develop a culture of some kind.

Homo sapiens sapiens. Finally, the emergence of fully modern humans, now called *Homo sapiens sapiens,* or doubly "wise" man, came about some 40,000 to 15,000 years ago. This creature was physically identical to today's humans and had real language. Humans moved from being strictly hunters and gatherers to domesticating animals and plants and creating fine artwork. Soon, their settlements had turned into real cities, and a civilization was created based on agriculture (farming). Recent studies of fossil fragments found in a cave in Israel suggest that modern humans (*Homo sapiens sapiens*) may have existed as far back as 92,000 years ago. Despite more information regularly coming to light, the human race still has not solved the complete mystery of human evolution at the beginning of the twenty-first century.

[*See also* **Evolution; Fossil; Genetics; Hominid;** *Homo sapiens neanderthalensis;* **Primates**]

Human Genome Project

The Human Genome Project is the scientific research effort to construct a complete map of all of the genes carried in human chromosomes. The finished blueprint of human genetic information will serve as a basic reference for research in human biology and will provide insights into the genetic basis of human disease.

The human "genome" is the word used to describe the complete collection of genes found in a single set of human chromosomes. It was in the early 1980s that medical and technical advances first suggested to biologists that a project was possible that would locate, identify, and find out what each of the 100,000 or so genes that make up the human body actually do. After investigations by two United States government agencies—the Department of Energy and the National Institutes of Health—the U. S. Congress voted to support a fifteen-year project, and on October 1, 1990, the Human Genome Project officially began. It was to be coordinated with the existing related programs in several other countries. The project's official goals are to identify each of the more than 100,000 genes in human deoxyribonucleic acid (DNA) and to determine the sequences of the 3,000,000,000 base pairs that make up human DNA. The project will also store this information in databases, develop tools for data analysis, and address any ethical, legal, and social issues that may arise.

In order to understand how mammoth an undertaking this ambitious project is, it is necessary to know how genetic instructions are carried on the human chromosome. Humans have forty-six chromosomes, which are coiled structures in the nucleus of a cell that carry DNA. DNA is the genetic material that contains the code for all living things, and it consists of two long chains or strands joined together by chemicals called bases, or nucleotides, all of which is coiled together into a twisted-ladder shape called a double helix. The bases are considered to be the "rungs" of the twisted ladder. These rungs are made up of only four different types of submolecules called nucleotides—adenine (A), thymine (T), guanine (G), and cytosine (C)—and are critical to understanding how nature stores and uses a genetic code. The four bases always form a "rung" in pairs, and they always pair up the same way. Scientists know that A always pairs with T, and G always pairs with C. Therefore, each DNA base is like a letter of the alphabet, and a sequence, or chain of, bases can be thought of as forming a certain message.

The human genome, which is the entire collection of genes found in a single set of chromosomes (or all the DNA in an organism), consists of 3,000,000,000 nucleotide pairs or bases. To get some idea about how much information is packed into a very tiny space, a single large gene may consist of tens of thousands of nucleotides or bases, and a single chromosome may contain as

The Human Genome Project has made great strides under the leadership of Francis Collins. In June 2000, researchers announced that they had completed a rough draft of the human genome. (Reproduced by permission of AP/Wide World Photos, Inc.)

FRANCIS SELLERS COLLINS

American geneticist Francis Collins (1950–) became the director of the National Human Genome Research Institute in 1993. The goal of this international scientific research effort is to construct a complete map of all of the genes (the units of heredity) carried on human chromosomes (a coiled structure in the nucleus of a cell that carries the cell's deoxyribonucleic acid). This finished blueprint of human genetic information will then serve as a basic reference for research in human biology and will provide insights into the genetic basis of human disease. Collins's own research is in the identification and understanding of genes that cause disease.

Francis Collins was raised on a small farm in Staunton, Virginia, where he was home-schooled until the sixth grade. After obtaining his bachelor's degree from the University of Virginia in 1970, he went on to obtain a doctorate in physical chemistry from Yale University in 1974. At this point, he became so intrigued by what he saw as the beginnings of a revolution in molecular biology (the study of the complex chemical molecules in all living things), that he switched fields and went back to school. After enrolling in medical school at the University of North Carolina, he began to study medical genetics and knew he was finally doing what he really wanted. After obtaining his medical degree in 1977, he returned to Yale for a fellowship in human genetics. It was there that he began working on the problem of trying to identify genes in human deoxyribonucleic acid (DNA) that cause disease. He continued studying this problem after joining the fac-

many as 100,000,000 nucleotide base pairs and 4,000 genes. What is most important about these pairs of bases is the particular order of the A's, T's, G's, and C's. Their order dictates whether an organism is a human being, a bumblebee, or an apple. Another way of looking at the size of the human genome present in each of our cells is to consider the following phone book analogy. If the DNA sequence of the human genome were compiled in books, 200 volumes the size of the Manhattan telephone book (1,000 pages) would be needed to hold it all. This would take 9.5 years to read aloud without stopping. In actuality, since the human genome is 3,000,000,000 base pairs long, it will take 3 gigabytes of computer data storage to hold it all.

In light of the project's main goal—to map the location of all the genes on every chromosome and to determine the exact sequence of nucleotides of the entire genome—two types of maps are being made. One of these is a physical map that measures the distance between two genes in terms of nucleotides. A very detailed physical map is needed before

ulty at the University of Michigan in 1984 where he developed an approach that is now called positional cloning.

Collins's approach was a major breakthrough in human genetic research as it allowed him to identify "disease genes" for almost any condition without knowing ahead of time what the particular abnormality might be. This approach has been adopted as a standard method of modern molecular genetics, and Collins demonstrated its usefulness in 1989 when his research team identified the gene for cystic fibrosis. The following year it did the same for the disease known as neurofibromatosis. This inherited disease causes tumors to form and destroys bones. In 1993 the gene for Huntington's disease was also located. That same year, Collins succeeded American biochemist James Dewey Watson (1928-), and became the second director of the National Center for Human Genome Research at the National Institutes of Health (NIH), heading the American part of an international project that involves at least eighteen countries. Collins also founded a new research program in genome research at NIH, and his own laboratory continued to focus on the identification and understanding of the genes that cause disease. Scheduled to be completed in 2003, the Human Genome Project is regarded by many as the most important scientific undertaking of our time, and Collins is bringing it to conclusion ahead of schedule and under budget. By the year 2000, the project had already compiled what might be called a rough draft of the human genome, having put together a sequence of about 90 percent of the total.

real sequencing can be done. Sequencing is the precise order of the nucleotides. The other map type is called a genetic linkage map and it measures the distance between two genes in terms of how frequently the genes are inherited together. This is important since the closer genes are to each other on a chromosome, the more likely they are to be inherited together.

As an international project involving at least eighteen countries, the Human Genome Project was able to make unexpected progress in its early years, and it revised its schedule in 1993 and again in 1998. Completion is expected in 2003, coinciding with the fiftieth anniversary of Watson and Crick's description of the molecular structure of DNA. During December 1999, an international team announced it had achieved a scientific milestone by compiling the entire code of an complete human chromosome for the first time. Another achievement was made in June 2000, when researchers announced that they had completed what might be called a rough draft of the human genome, having put together a sequence of about 90 percent of the total. Researchers chose chromosome twenty-two because of its relatively small size and its link to many major diseases. The sequence they compiled is over 23,000,000 letters in length and is the longest continuous stretch of DNA ever deciphered and assembled. What was described as the "text" of one chapter of the twenty-three-volume human genetic instruction book has been completed. Francis Collins, Director of the National Human Genome Research Institute of the National Institutes of Health said, "To see the entire sequence of a human chromosome for the first time is like seeing an ocean liner emerge out of the fog, when all you've ever seen before were rowboats." Another major advance on the project was made in June 2000 when researchers announced that they had completed a rough draft of the fully mapped human genome.

With this fully mapped genome, biologists can for the first time stand back and look at each chromosome as well as the entire human blueprint. They will start to understand how a chromosome is organized, where the genes are located, how they express themselves, how they are duplicated and inherited, and how disease-causing mutations occur. This could lead to the development of new therapies for diseases thought to be incurable as well as to new ways of manipulating DNA. It also could lead to testing people for "undesirable" genes. However, such a statement leads to all sorts of potential dangers involving ethical and legal matters. Fortunately, such issues have been considered from the beginning, and part of the project's goal is to address these difficult issues of privacy and responsibility, and to use the completely mapped and fully sequenced genome to everyone's benefit.

[*See also* **Chromosomes; Genes**]

Human Reproduction

Human reproduction is essential for the continuance of the human species. Humans reproduce sexually by the uniting of the female and male sex cells. Although the reproductive systems of the male and female are different, they are structured to function together to achieve internal fertilization.

It is a characteristic of all living things on Earth that they reproduce or produce offspring, and humans are no different. If humans are examined as large, complex land mammals, then we can say from a strictly biological point of view that the male and female have the same role as other mammals. The male's job is to produce sperm cells and deliver them into the female reproductive tract. The female's job is to produce ova (eggs), receive the sperm, and nourish the embryo that grows inside her. She must also give birth and produce milk for the offspring during its early years.

Unlike other land mammals, humans are not simply physical creatures who must follow the course nature has planned for them. People have intelligence and the ability to do what they want. People have invented culture and civilization, and have made up rules and values. A good example of how different humans are from all other living things is reproduction itself. Humans are the only ones who can choose not to reproduce, for whatever reason. So, despite the millions of years of evolution and adaptation that thousands of generations went through to produce a certain human being, that person can decide not to reproduce with another human and therefore not to pass on its genetic inheritance. Although humans are greatly influenced by their biological makeup, they are not compelled by it they way other animals are.

MALE REPRODUCTIVE SYSTEM

Among humans who do choose to reproduce (and this is by far the greater part of the human species) reproduction is basically a biological process. As mammals, humans practice internal fertilization, which means that the sperm and egg come together inside the female's body. In order for this to happen, both male and female need a set of organs and systems that work together. The male reproductive system produces, stores, and releases its gametes, or sex cells, known as spermatozoa. Sperm are produced in the testes, two oval-shaped organs contained in the scrotum, which is a like a pouch of skin. Located outside the body, it keeps sperm a few degrees lower than 98.6°F (37°C), since normal body temperature would kill most sperm. The testes are made of tightly coiled tubes in which sperm cells are formed. They are stored internally in a liquid called seminal fluid that keeps them nourished. The penis is the external part of the male reproductive system and contains a central channel called the urethra. Sperm flow out of the urethra at the proper time. As the specialized organ through which sperm is introduced into the female reproductive tract, the penis is made of spongy tissue that lengthens and stiffens when excited. At this point, it is ready and able to be inserted into the female's vagina.

FEMALE REPRODUCTIVE SYSTEM

The female gametes, or sex cells, are known as eggs, or ova, and are produced and stored in the ovaries. An egg is 75,000 times larger than an individual sperm cell. Females are born with about 40,000 immature eggs and do not produce any more during their lifetime. In most females, only about 400 of these eggs actually mature. From the onset of puberty until sometime in their forties, females release one mature ovum approximately every month. This monthly release of an egg that is ready to be fertilized is part of the female's menstrual cycle. The term menstrual comes from the Latin word *mensis,* which means "month." Therefore, every twenty-eight days an egg matures and is positioned to meet with a sperm cell in the Fallopian tubes. These two, 3-inch (7.6-centimeter) tubes connect the ovaries with the uterus. It is in the Fallopian tubes that fertilization takes place.

FERTILIZATION

The uterus is a muscular structure that houses the developing egg if it is fertilized. The vagina is the muscular tube leading from the uterus to the outside of the body and it is the entrance or canal through which the male deposits his sperm. The act of human sexual reproduction is called sexual intercourse or coitus, and for it to work properly, both partners must usually be excited or experience what is called sexual arousal. This sexual stimulation results in body changes, such as the male's erect penis and the female's lubricated vagina. Nature has arranged it so that sexual reproduction is pleasurable or feels good to most organisms, and for humans, the inward and outward movement of the penis in the vagina

An ultrasound of a nine-month-old human fetus. A woman's reproductive system is designed to be able to carry and protect an unborn child. (Reproduced by permission of Photo Researchers, Inc. Photograph by Matt Meadows.)

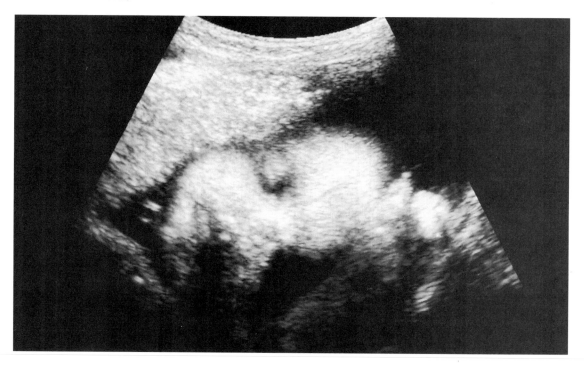

causes friction that leads the male to an orgasm. In the male, orgasm causes very strong, involuntary contractions of muscles at the base of the penis. These contractions forcefully expel semen, which contains sperm, from the penis. This release of sperm is called ejaculation. The female may or may not experience a similarly intense feeling, but even if she does not, she can still become pregnant.

Hundreds of millions of sperm cells are released during an ejaculation, and they swim through the uterus and into the Fallopian tubes. Many sperm attach themselves to the egg but only one actually enters it. Once a sperm enters the egg, the egg prevents any others from doing so. If fertilization occurs—and for many reasons it often does not—the zygote (fertilized egg) begins to divide and grow. It will then implant itself in the uterus where it will be nourished and grow into a baby. If the egg was not fertilized, it is eventually discharged out of the vagina with other uterine tissues and blood. This is sometimes called a woman's "period."

A fertilized egg that successfully attaches to the uterus will take about 270 days to grow into a fully developed fetus or baby. When it is ready to be born, the baby's adrenal glands secrete a hormone that signals the mother's pituitary gland to secrete a hormone called oxytocin. This causes the uterine muscles to contract rhythmically, and eventually the baby is born, or expelled, from the uterus.

Hybrid

A hybrid is the offspring produced by organisms of two different varieties or species. Hybridization occurs often in nature between different varieties of the same species, but much less often between related or different species. The product of such a cross is usually unable to reproduce itself.

A hybrid is basically the product of two different organisms. In the plant kingdom, hybrids occur all the time both by natural pollination and when humans deliberately cross different types. Wheat is a hybrid that came about naturally, although the types grown now are hybridized even further by farmers so they will resist certain diseases and produce higher yields. Probably the most famous hybridizer was the Austrian monk and botanist (a person specializing in the study of plants) Gregor Johann Mendel (1822–1884), whose famous experiments with garden peas led to his discovery of the laws of heredity. Mendel spent years deliberately crossing different varieties within a species to produce other new varieties. Mendel started by crossing plants that bred different true traits (such as all tall or all dwarf), and produced hybrid plants whose varied offspring eventually led him to discover the phenomena he called dominant and recessive traits. Today, different types of plants are crossbred by farmers to produce a particular combination of desirable features.

In the animal world, a hybrid is more of an exception. Nature seems to work against members of different species trying to mate and reproduce, and if fertilization does somehow occur, the result is usually not able to survive much past birth. Very often, the fertilized egg does not develop properly and it dies. In instances where the two animals are members of different but closely related species, the offspring is born but it is usually sterile or unable to reproduce. The best-known animal hybrid is probably the mule. In this case, two closely related species, the horse and the donkey, are able to mate and produce an offspring. However, because the horse has sixty-two chromosomes and the donkey has sixty-four, the hybrid mule is born with sixty-three. It is therefore unable to successfully fertilize another animal because its odd-numbered chromosomes are unable to pair up correctly during meiosis (the special form of cell division that produces sex cells). A mule is a cross between a female horse and a male donkey. A "hinny" (also sterile) is the result of a male horse mating with a female donkey. While both mules and hinnies can be healthy, vigorous animals, they are nonetheless unable to reproduce because of what is called hybrid infertility. As a result, self-sustaining mule or hinny population could never develop since both must necessarily be produced by repeating the original crossbreeding.

[*See also* **Fertilization; Genetics**]

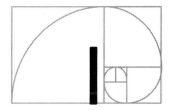

Ichthyology

Ichthyology is the branch of zoology (the study of animals) that deals with fish. It includes the study of the development, anatomy (structure), physiology (function), behavior, classification, genetics, and ecology of fish, among other things. Since fish are a major food source for people, the study of ichthyology also has economic importance.

Taken from the Greek word *ichthys* for fish, ichthyology had its beginning with the Greek philosopher Aristotle (384–32 B.C.), since the ancient world was more interested in and more knowledgeable about fish than they were about many other animals. This may have been because fish were both a relatively easy-to-obtain source of food as well as an animal group that was readily accessible, since fishing is one of humankind's oldest occupations. Until the end of the nineteenth century, however, more attention was paid to describing and classifying fish than any other aspect. By then, ichthyology was well on its way to becoming a separate field of zoology. By the end of the first half of the twentieth century, the emergence of oceanography (the science of the ocean) and the newfound ability to conduct underwater observations, allowed scientists to be able to study fish in their natural environment for the first time. The development of improved techniques for keeping fish in tanks for study also spurred further advances.

There are more than 22,000 known species of fish in the world, and they live in nearly every imaginable body of water, from stagnant ponds to the deepest oceans. They live in water all of the time and breathe through gills. Together with mammals, birds, reptiles, and amphibians, fish are one of the major groups of vertebrates (animals with a backbone).

They are considered the most successful vertebrate group, outnumbering birds two to one and mammals seven to one. An ichthyologist, therefore, must contend with a great variety of subjects, from the bony, snakelike eel and the shark with all cartilage and no bone, to the bioluminescent deep-sea fish that can make its own light.

Today, fish are not studied just for their own sake or to simply learn more about them. Since fish are a major food source and fishing is an important industry, a great deal of fishery research is conducted in government laboratories as well as institutional aquariums. It is not surprising that much of this work is aimed at learning more about diseases in fish as well as understanding the effect that pollution has on them. Fishes are as vulnerable to infections by viruses as are higher vertebrates, and often they are the first to show signs of disease. They are also susceptible to tumors, and sick fish are a signal that they live in an environmentally degraded body of water. Increasingly, ichthyologists must know as much about the environment of a certain type of fish as they do about the fish itself in order to note any irregularities in the environment—and thus in the fish.

[*See also* **Fish**]

Immune system

The immune system is the body's biological defense mechanism and protects it against foreign invaders, such as bacteria and viruses. The system is a collection of cells and tissues in the body that protect it against disease-causing organisms. It works by using a simple system in which it distinguishes self (acceptable) from nonself (nonacceptable), and then it attacks and attempts to destroy anything nonself. Nearly all animals, simple and complex, have an immune system based on this self/nonself mechanism.

NATURAL IMMUNITY

The immune system has two different types of defense. The first is called natural immunity and is composed of the basic physical and chemical barriers that every body has at its disposal to fight a foreign invasion. The body's skin is its first line of natural defense since healthy, unbroken skin acts as a physical barrier against microorganisms. If, however, these tiny organisms try to get into the body through normal openings, like the nose and eyes, the body is prepared. These passages are lined with sticky mucus that catches the microorganisms, and with hairlike cilia that sweep them back out of the body. The body also uses secretions, like tears and saliva to protect itself. These secretions contain an enzyme called lysozyme that breaks down the walls of invading bacteria. If after these three defenses, microorganisms still manage to get into the body, the blood contains certain types of white cells called phagocytes that literally swallow up and destroy foreign cells or substances. The body then activates its complement system, which releases proteins that cause an inflammatory response. With this response, the body releases a fluid called histamine that helps fight the invader and results in local swelling.

ACQUIRED IMMUNITY

Sometimes the phagocytes, which make a general attack and attempt to destroy anything detected as foreign, cannot cope with the invader. When this happens, the body's acquired immunity goes into action. If natural immunity is the body's "nonspecific defense," meaning it will attack anything detected as foreign, then acquired immunity is its "specific defense."

Acquired immunity allows the body to "remember" and link past infections to a particular bacteria or virus. The body is then able to respond more quickly the next time it encounters the invader (now called an antigen). Only vertebrates (animals with a backbone) have acquired immunity. Acquired immunity enables the immune system to produce certain types of white cells called antibodies to fight a particular type of pathogen (disease-producing organism). It also enables the immune system to "remember" that pathogen and to respond more quickly the next time it appears. The body produces three types of white blood cells, macrophages, T lymphocytes, and B lymphocytes, that work together and carry out a complex series of events known as the immune response. (Macrophages alert the immune system that specific foreign agents are present.) The primary immune response involves the B-cell lymphocytes producing antibodies that capture and kill the invading antigen. However, when a virus invades a cell, the virus is safe from antibodies, so the T-cell lymphocytes begin what is called cell-mediated immunity. The T-cell is able to recognize any infected cell, and when it does, it kills the cell. Lymphocytes are transported throughout the body via the lymphatic system, a type of secondary circulatory system that acts as a bridge to the immune system.

EDWARD JENNER DEVELOPS IMMUNIZATION

The natural ability of the immune system to be able to recognize a particular antigen is the basis for immunization. Since ancient times, medical observers had noticed that the body seemed to have powers to protect itself and resist disease. In particular, people who had survived a certain infectious disease did not suffer from that disease again during their

PAUL EHRLICH

German bacteriologist (a person specializing in the study of bacteria) Paul Ehrlich (1854-1915) is the founder of chemotherapy, which is the use of a chemical substance to treat a disease. He also identified substances that could be used as drugs to destroy bacteria in the body, and made important contributions to the understanding of immunity (the body's natural resistance to a foreign substance).

Paul Ehrlich was born in Strehlin, Silesia (then part of Germany, now part of Poland). His family was educated and well-off, and although young Ehrlich did not do well in school at first, he came to be very interested in both chemistry and biology. He attended German universities and received his medical degree in 1878. Throughout his medical education, Ehrlich was always interested in its chemical aspects, and he became especially interested in the new dyes that were being introduced. Ehrlich was particularly fascinated by the staining (dyeing) of cells and tissues and their reactions to dyes. For his graduate thesis he discovered several practical stains for bacteria and even wrote his thesis on that subject. After working with the famous German bacteriologist Robert Koch (1843-1910) studying tuberculosis, he was appointed a professor at the University of Berlin in 1890.

There he began work with others on the study of immunity, or the body's own defense against disease. The group he joined was trying to find a cure for diphtheria, a childhood respiratory disease that killed many. Ehrlich was searching for a substance that would give immunity against diphthe-

lifetime. In 1796, the English physician Edward Jenner (1749–1823) discovered that it was possible to make people immune to a disease they never had. First, he gave a person an injection of a dead or weakened microorganism (called a vaccine) that caused a certain disease (like cowpox). The vaccine was not strong enough to give the person cowpox, but still the patient's body would react by producing antibodies against the disease. Jenner found that immunization protected his patients from the dreaded smallpox disease. Eventually, successful methods of immunization were developed against such diseases as diphtheria, whooping cough, mumps, measles, rubella, polio, rabies, anthrax, typhoid fever, typhus, yellow fever, cholera, and the plague.

HIV AND AIDS

The last few decades of the twentieth century witnessed not a new disease to fight, but the emergence of a disorder of the human immune

ria by using antitoxins. Antitoxins are antibodies produced by the body's immune system to fight poisons invading the body. An antibody is a special protein in the blood that locks on to a specific foreign substance and kills it. By 1892, Ehrlich had worked out an antitoxin for diphtheria that could be used medically. He obtained the right antitoxin from large animals that had been immunized against diphtheria. He then concentrated and purified it and administered it to 220 children with success. For his work on immunity, Ehrlich later won the 1908 Nobel Prize for Physiology and Medicine. After this achievement, Ehrlich returned to studying dyes and stains, and decided to pursue a fascinating idea. He knew that stains were useful because they colored some cells but not others, thereby making the stained ones stand out. He also knew that a stain would not color a bacterium (plural, bacteria) unless it combined with something in the bacterium. Knowing also that when this happened the bacterium usually died, he theorized that if he could find a dye that stained bacteria but not ordinary cells, then maybe it was also a chemical that killed bacteria without harming the host (the human being). He described such a chemical as a "magic bullet," saying that it would seek and destroy only the invader. Eventually, he did discover one dye, called trypan red, that worked against such diseases as sleeping sickness. Much later, he discovered a dye he named Salvarsan that would kill the microorganism that caused syphilis, a sexually transmitted disease. These two chemicals marked the beginning of modern chemotherapy. Ehrlich proved to be a pioneer not only in the field of immunology, but in the newer field of chemotherapy as well.

system itself called AIDS (Acquired Immune Deficiency Syndrome). Infection by this new Human Immunodeficiency Virus (HIV) caused the immune system to collapse, leaving the body defenseless. Specifically, the HIV virus attacked certain T-cells and made them unable to do their job helping B-cells make antibodies. The result was that once a person's natural immune system shut down, they became host to a number of devastating infectious organisms. AIDS is not a single disease but a syndrome of symptoms that are caused by infectious invaders taking advantage of an immune system that cannot function. HIV can remain dormant in the body for some time without producing any signs. There is still no cure for AIDS, although great progress has been made in coping and managing this disease and prolonging the lives of its victims.

Also in the last few decades, biologists have discovered that the immune system can be affected by a person's psychological health or state of mind. Apparently there exists a complex network of nerves, hormones, and brain chemicals that link the immune system to a person's mental state, and it has been demonstrated that extreme psychological stress can suppress the immune system and accelerate certain diseases. Recent immunological research indicates that the mind/body connection is more significant than previously thought.

[*See also* **AIDS; Antibody and Antigen; Immunization; Lymphatic System; Vaccine**]

Immunization

Immunization is a method of helping the body's natural immune system be able to resist a particular disease. It is usually carried out by giving someone a mild version of the disease. This allows the body to make antibodies that will resist the disease in the future. Active immunization, or vaccination, has proven to be a highly successful method of disease prevention.

Long before modern science discovered the causes of disease, it was folk practice in some parts of the world to give a powder made from the scabs of recovering smallpox patients to healthy children in the belief that it would somehow protect them in the future. If this risky custom, which originated in China, did not kill the child, it often did grant him or her immunity against a full-blown case of smallpox (a highly infectious viral disease). This same idea was at work when the English physician Edward Jenner (1749–1823) decided to try a dangerous experiment. He based his experiment on the fact that people who had suffered a case of the less serious cowpox (a contagious skin disease found in cattle) often did not catch the deadly smallpox. In 1796 Jenner prepared what he called a vaccine (because the cowpox virus name was "vaccinia") and gave it to a young boy. Months later, he injected real smallpox into the boy. Fortunately, the boy did not get the disease.

This marked the modern beginnings of immunization. Jenner, however, made no claims that he understood why immunization worked. It was not until a century later that the French chemist and microbiologist (a person specializing in the study of microorganisms) Louis Pasteur (1822–1895) proved experimentally that disease-causing microorganisms (organisms that can only be seen through a microscope) that were "attenuated," or weakened, would create an immune response in a person without actually causing the disease itself. On the basis of this breakthrough, active immunization began. It is now known that immunization uses the mechanisms of the body's natural immune system to protect the body against future diseases.

IMMUNIZATION AND ANTIBODIES

In the twentieth century, science learned that the body produces substances called antibodies. These fight and kill what the body recognizes as foreign invaders (disease-producing microorganisms). These antibodies are specific to that particular disease and will remain in the body's "memory cells" for a long time, ever ready to fight should the body recognize the disease in the future.

Today, many viral vaccines are made from live, weakened viruses, including those for yellow fever, measles, mumps, rubella, and polio. Using a live form means that the body will react with a very strong immune response to the particular virus, thereby protecting the individual against future infection. Other vaccines, like rabies, flu, and intravenous polio, use dead viruses and do not confer as strong a protection.

PASSIVE IMMUNIZATION

All of the above are considered to be forms of active immunization, but there is also another method called passive immunization. This method is used when a quick response to a disease is required. Passive immunization consists of injecting specific antibodies into a person to fight a specific disease. For example, a person who is bitten by a snake or who has been exposed to hepatitis cannot wait for his own system to build up antibodies against them. Instead, the person is given a direct and immediate dose of the antibodies in order to neutralize the venom or to kill the microorganism. While passive immunization usually works, it is not long-lasting like active immunization and usually will not protect the person in the future.

IMMUNIZATION WORKS

Immunization has proven to be the safest, least expensive, and most effective means of protecting people

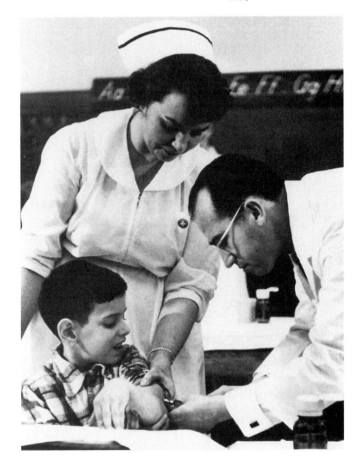

Doctor Jonas Salk, a pioneer in immunization, immunizing a child against polio. Immunization has proved to be the best way to protect people against contagious diseases. (Reproduced by permission of AP/Wide World Photo, Inc.)

against contagious diseases. In many ways it is ideal, since it prevents the disease rather than trying to cure it once it has taken hold. Today, immunization in the developed world has nearly eliminated the threat of typical childhood diseases like measles, mumps, rubella, whooping cough, and polio. In 1980 the World Health Organization declared that smallpox was the first disease to be totally eradicated (destroyed) worldwide.

[*See also* **Antibody and Antigen; Immune System; Vaccine**]

Inbreeding

Inbreeding is the mating of organisms that are closely related or that share a common ancestry. It is used deliberately by people to try to retain desirable traits and eliminate undesirable ones in animals. However, inbreeding can result in harmful recessive genes that had been masked in parents but later appear in the offspring.

Inbreeding of animals has been conducted by people ever since they first began to keep animals for food, clothing, and transport. It probably was done initially when a particularly useful characteristic was displayed in an animal, and the animal was then encouraged to mate with another of its kind that had the same, desirable trait. Today, inbreeding is used in animal husbandry, which is the scientific control and management of animals. It is performed for the same original purpose of encouraging the development of certain desirable traits in offspring. A good example of inbreeding is done with dogs and cats who are bred primarily for their appearance. There are also some types of inbreeding that occur in nature. Self-fertilization is one example that occurs in bisexual flowering plants. This is probably the most "inbred" an organism can be since its offspring are the result of the fusion of the male and female sex cells of the same individual. This particular form of inbreeding is sometimes necessary since it allows an isolated individual plant to create a local population. One disadvantage of such an inbred population, however, is that its ability to adapt to environmental changes is limited since its members all share the same pool of genes.

This limitation of inbreeding applies to animals as well, and all farmers know that they can only mate animal siblings (brothers and sisters) for a few generations before they start to show signs of being less healthy and less fertile. The reason for this is because of the same lack of variability (genetic differences) that the isolated flower population suffered. Inbreeding can cause harmful genes that are recessive in both parents to become expressed in their offspring. A recessive gene is not expressed if there is a dominant one to offset it or mask it. However, it is still part of the individual's genetic makeup. The recessive gene will be expressed if the other parent also has a recessive gene for the same trait.

Continued inbreeding can result in an accumulation of recessive genes, which can cause what is called "inbreeding depression." This is not a state of mind but rather a physical state that results in low fertility, poor general health, and particularly negative characteristics like stunted growth or feebleness. One example of a species that is threatened by its lack of genetic diversity is the cheetah. Because today's existing cheetahs are all so closely related, most have weakened immune systems and are very susceptible to disease. To avoid this, agricultural breeders sometimes practice a form of inbreeding called "linebreeding." This is accomplished by mating a female animal with its grandfather or uncle (rather than with a sibling or parent). This reduces the probability of undesirable genes in the offspring.

The opposite of inbreeding is outbreeding, which is defined as mating individuals that are not related at all. While this can make it more difficult to regularly achieve a certain desirable quality or trait, it does produce more vigorous and healthy offspring. The most extreme example of outbreeding is called crossbreeding in which individuals of different but closely related species are mated. The mule is an example of the crossbreeding of a horse and a donkey, and although it is a very strong and useful animal, the mule is nonetheless sterile or cannot reproduce. This is the case with all crossbred animals.

[See also **Genetic Disorders**]

Inherited Traits

An inherited trait is a feature or characteristic of an organism that has been passed on to it in its genes. This transmission of parental traits to their offspring always follows certain principles or laws. The study of how inherited traits are passed on is called genetics.

The study of genetics or heredity began in the early 1800s when scientists first began trying to explain the existence of different species and variations within the same species. The French naturalist, Jean Baptiste de Lamarck (1744–1829), was one of the first to seriously consider the idea that present-day life forms descended from common ancestors. This is the principal idea of biological evolution. However, Lamarck also put forth some incorrect notions about biological development and heredity. One of these was that new organs and capabilities could be developed out of need and would also grow and improve when routinely used over time.

GREGOR JOHANN MENDEL

Austrian botanist (a person specializing in the study of plants) Gregor Mendel (1822-1884) is considered to be the father of genetics. After years of breeding peas and studying their characteristics, he discovered the basic laws of heredity that apply to all plants and animals. His work not only explained English naturalist Charles Darwin's theory of evolution (the process by which living things change over generations) by natural selection, but laid the foundation of modern genetics.

Gregor Mendel was named Johann when he was born in Heinzendorf, Austria (now part of the Czech Republic). The son of a peasant who took care of the fruit trees on a rich man's estate, the young Mendel took the name Gregor when he became a priest. Although very poor, he had been helped by the church to obtain a basic education and eventually received some higher training in mathematics and science. However, when his financial situation got very bad, he entered a monastery in 1843, mainly as a means of trying to continue his education. Although he had not intended on becoming a priest when he entered the monastery, four years later he decided to become a priest and was ordained that year. Eventually he was sent to the University of Vienna to study zoology, botany, chemistry, and physics. After becoming a science teacher, he repeatedly failed the examination that would have enabled him to teach at a higher level, so he finally just gave up and decided to pursue his own interests while remaining a priest at the monastery. Since he was particularly interested in both mathematics and botany, he decided to combine his two loves and to see if it were possible to predict the kinds of fruits and flowers a plant would produce.

Until Mendel, no one had ever done any real statistical analysis of breeding experiments. So starting in 1868, Mendel began a long-term project—almost a hobby—to see if he could conduct a range of breeding experiments and keep accurate records of his results. Mendel wanted to see if he could begin to understand how traits pass from one generation to another, so after taking over the monastery's garden plot, he chose to breed pea plants (today's "sweet peas"). Peas were an especially good experimental plant be-

In other words, Lamarck said that a characteristic could be *acquired* (the long neck of a giraffe was acquired by continuously stretching its neck to reach for leaves) and then could be passed on to its offspring. He also said the opposite, that those characteristics that were not used would eventually disappear. Lamarck also argued that when an organism acquired a new skill it passed on that ability to its offspring. Of course, we know today that if one person learns a foreign language, there is no *genetic* way he or she can pass that talent on to its children. However, Lamarck was

cause they had characteristics, or traits, that could be easily observed (tall or short, wrinkled seeds or smooth seeds, yellow or green seeds). Mendel was very careful in all of his experiments, transferring pollen by hand from the male to the female part of the flowers to produce seeds. He even wrapped his plants so they would not be accidentally pollinated by insects. He would save the seeds from each self-pollinated plant, plant them separately, and study the new generation. He also crossbred plants with different characteristics.

After eight years of work and careful recording, Mendel found that, indeed, he was able to predict what a certain plant would produce as long as he knew which plants were the parents. In fact, he was so certain of what he found that his conclusions are now called Mendel's "laws" of inheritance. What he discovered after years of breeding more than 30,000 plants was that there are powerful traits, called dominant, and weaker traits, called recessive. He also found that when mixed together they do not blend. For example, although he at first expected to breed a medium-size plant when he crossed a tall plant (dominant trait) with a dwarf plant (recessive trait), what he found was that he eventually wound up with a mixture of tall and short plants according to a given ratio. He concluded, therefore, that in every instance, mixing traits did not result in a blend but instead sorted themselves out according to a fixed ratio. Mendel also concluded that each parent plant contributed a factor, later found to be a gene, that determines what a certain trait will be. Unfortunately for Mendel, he published his results in a journal not read by many in Europe. When he wrote directly to a prominent botanist of his time, the heavily statistical arguments he offered confused a man who was unused to seeing mathematical data in a botany paper. Discouraged, Mendel later put away his work and died totally unnoticed. It was not until 1900 that his work was discovered and made public. Upon close consideration, a new generation of life scientists realized that Mendel's laws of inheritance supported and even explained Darwin's theory of evolution by natural selection. Thus, the quiet priest who worked with the humble garden pea is now recognized as the giant who laid the groundwork for the modern science of genetics.

on the right track since he did suggest that traits can be inherited from generation to generation and that species do undergo long-term evolutionary changes.

In 1859, the English naturalist Charles Robert Darwin (1809–1882) published his landmark work, *On the Origin of Species,* in which he outlined his theory of evolution through natural selection. Darwin argued that members of a particular species always have slightly different traits or characteristics, and that in the competition for food, space, and shelter, some of these differences would make one member more suited to survive and produce offspring than others of its species. He continued this line of reasoning and said that those traits that were an advantage would be passed on to later generations, while those that were not would eventually disappear as their carriers died out. This meant that after centuries upon centuries of competition, or natural selection, recent members of a species could be quite different from their ancestors. Despite its eventual acceptance by the scientific community, Darwin's theory lacked an explanation for the mechanism or manner in which these random variations were inherited.

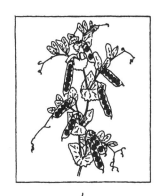

It was not until 1900 that the means of transmission of inherited traits was understood. Some forty-five years earlier, the Austrian monk, Gregor Johann Mendel (1822–1884), had begun experimenting with pea plants at about the same time that Darwin set forth his ideas on natural selection. Through his careful experiments, Mendel demonstrated that what he called "hereditary factors," now called genes, are transmitted to offspring. He also discovered that traits are inherited in pairs and that usually only one trait in each pair is actually expressed in the offspring. Although Mendel had established the laws of heredity by 1865, his work remained unnoticed until it was independently rediscovered by three scientists in 1900. With their discovery, it became apparent that Mendel had formulated the fixed rules of inheritance that applied to the entire plant and animal kingdoms.

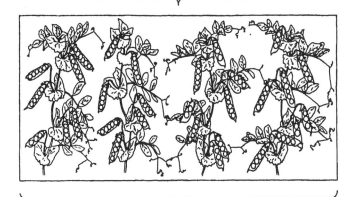

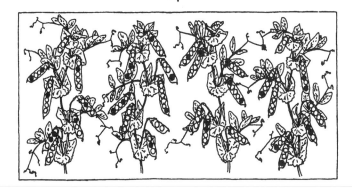

After 1900, science began its search for the key part in all living things that contained the actual information that determined every detail of what an organism was. This search soon led to an understanding of chromosomes (a coiled structure in the nucleus of a cell that carries the cell's heredity information), and then to the realization that they in turn were made up of other, smaller things later named "genes." By the early 1950s, science knew that the chemical deoxyribonucleic acid (DNA) was somehow at the center of heredity, and in 1953 the American biochemist, James Dewey Watson, and the English biochemist, Francis Harry Compton Crick, explained exactly how. That year they discovered the "double helix" structure of the DNA molecule, demonstrating exactly how DNA carries the genetic code for all living things.

By the end of the twentieth century, our knowledge of the mechanism of inherited traits had nearly reached the point where the actual location of every human gene on every chromosome was identified and every letter in the 3,000,000,000-base code deciphered. Finally, on June 26, 2000, scientists announced that they had completed a rough draft of the human genome—the complete set of chromosomes that determines humans inherited traits. When completed, this human genome will lead to an understanding of each gene's precise chemical structure and its function in health and disease. This information will be invaluable since it may lead to cures, or possibly preventions, of certain genetic disorders.

[*See also* **Dominant and Recessive Traits; Genes; Genetics; Mendelian Laws of Inheritance**]

Insects

An insect is an invertebrate animal with six legs and a body that is clearly divided into three main segments. The heads of most insects have a pair of antennae, compound eyes, and large jaws. Insects inhabit nearly every part of Earth and make up the most numerous class of living animal. They are considered to be the most successful group of living creatures ever to have lived on Earth.

Insects belong to the phylum Arthropoda and make up its largest class. Well-known examples include bees, ants, butterflies, grasshoppers, beetles, moths, mosquitos, and the house fly. Insects are so diverse that there is probably no typical insect, although there is a basic insect anatomy or structure. An adult insect has three distinct body segments—the head, thorax, and abdomen. The head contains the sense organs like antennae and eyes, as well as three pairs of mouthparts that are adapted for either biting, chewing, puncturing, or sucking, depending on the species. The thorax has three segments, each of which has a pair of legs that are used for

Opposite: A nineteenth-century diagram illustrating Gregor Mendel's discovery of the patterns of inheritance as shown by sweet peas. This diagram shows the original crossing, the first generation, and the next when recessive traits appear in the proportions discovered by Mendel. (Reproduced by permission of The Granger Collection Ltd.)

walking and clinging. If an insect has wings, they are attached at the thorax. Also located at the thorax are small tubelike openings called trachea that insects use to take in oxygen and expel carbon dioxide. Some of these openings are also found on the forward part of the abdomen. Since insect tissue gets oxygen directly through these tracheae, the circulatory system is fairly simple. The insect abdomen is used primarily for reproductive purposes. Digestion occurs in a three-part gut and wastes are excreted out of very specialized organs called Malpighian tubules. Named after the Italian anatomist (a person specializing in the structure of animals) who discovered them, these function like kidneys and remove waste from the insect's system. Insects also possess an exoskeleton, a hard outer support structure that protects their soft internal organs and provides some protection against predators.

Insects have been described as the dominant form of life on Earth. One writer even argues that if visitors from other planets came to Earth and studied its life forms, they might want to communicate first with an

A labeled diagram showing the external and internal features of a generalized insect. (Illustration by Hans & Cassidy. Courtesy of Gale Research.)

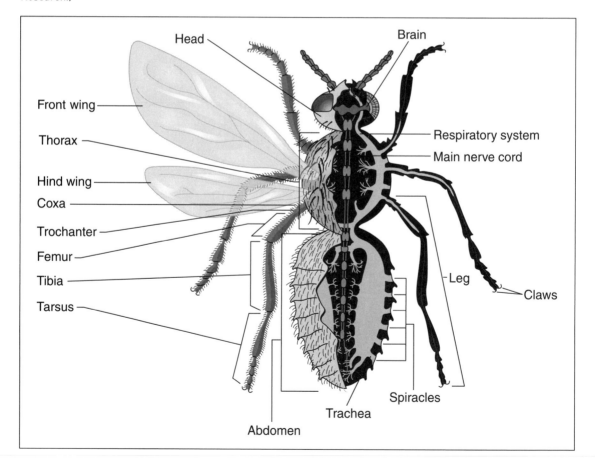

insect because of their staggering diversity and overwhelming numbers that the insect represents. One reason for this steady success is the small size of insects. A small insect does not need as much food as a large animal does, so it is easier for an insect to find food. Insects also have highly specialized mouthparts and digestive systems that allow them to consume almost any plant in existence. If necessary, however, they will eat anything they can find. An insect's small size also allows it to hide more easily from its enemies, thus possibly enabling the insect to avoid predators and live longer.

Another reason for their success could be the distinct advantage having wings provide insects. Being able to fly means insects can easily escape earthbound predators, but it also confers the advantage of being able to leave a certain habitat if it proves difficult or dangerous to live there. Flying also allows insects to populate other habitats that most animals would have a hard time reaching. Altogether, their small size and flying ability gives insects a competitive advantage in the struggle for survival.

Another major reason that certain insects prosper is because they are social insects. Although some live independently of others, some insects live together in what are called colonies. A colony is a group of animals that live together and share work and food. Honeybees, termites, ants, and wasps are social insects. In a colony, an individual insect does not have to provide itself with food and shelter, since all the tasks needed for living are carried out by different members of the colony. Each member has a specific job to do, and it performs only that role and nothing else. For example, in an ant colony, a queen ant lays eggs and contributes to the growth of the colony; soldier ants guard the colony and do the fighting during an attack; nurse ants tend only to the queen, her eggs, and the larvae; and worker ants locate food outside the colony and bring it back for all. All of the roles are performed totally by instinct. The system works because each insect in a colony spends every day of its life doing the job it must do and nothing else. Besides being social, insects also have a staggering reproductive capacity. For example, if all the eggs of a single fly were to survive and reproduce through only six more generations, there would be more than 5,000,000,000,000 flies.

In many ways, insects are mankind's most aggressive competitor. They will eat crops as well as stored food. They can swarm and consume every green thing in sight. They also can destroy paper, wood, and cloth. They bite humans and other animals and transmit diseases.

On the other hand, insects are important and necessary pollinators of flowers and crops, and many beneficial insects (like the ladybug and praying mantis) attack or destroy many insects harmful to humans or crops.

Bees provide honey as well as pollination. Overall, human beings have now learned that we cannot defeat or even diminish the range, extent, and diversity of the insect population. However, humans can control insects' negative effects by learning more about their habits, needs, and life cycles.

[*See also* **Invertebrate**]

Instinct

Instinct is a specific inborn behavior pattern that is inherited by all animal species. Instinctive behavior exists at birth and does not have to be learned. Most instinctive behavior is related to an animal's survival.

The word instinct could be used to describe the set of wired instructions that are built into an animal's nervous system. These instructions are passed genetically from one generation to the next. From observing animal behavior, it is known that particular species will do certain things automatically almost from the moment of birth. For example, newborn chicks "instinctively" open their mouths when an adult bird returns to the nest. A baby kangaroo rat instantly performs an escape jump maneuver when it hears the sound of a striking rattlesnake, even if it has never seen a snake before. Nest-building and web-spinning also are examples of instinctive behavior that can be seen and observed.

All of these and other instinctive behavior patterns have things in common. Given the proper condition, situation, or stimulus, instinctive behavior patterns are automatic and are performed in a fixed, regular way by each member of the species. Each cocoon-spinning spider builds its silk cocoon in a certain sequence. The spider first spins a base plate, then the walls, lays its eggs, and adds a lid. The spider can only build the cocoon this way. If interrupted and moved elsewhere after having built the base plate, it will not start over but will continue as if the plate were already there. This means that the eggs will fall out the bottom when laid. Another feature common to all instinctive behavior is that it requires no learning and is carried out fully and completely the first time it is performed.

SCIENTISTS TROUBLED BY INSTINCT

Although the term instinct is commonly understood by most people, for scientists it presents a problem. The word does little to explain the real mechanisms that are at work. If instinct is used to describe behavior that is performed as if guided by some mysterious and unknown force, then "instinct" is not a real scientific word, since science seeks explanations. Many scientists do not use the word "instinct" anymore, since this is too general a word. Instead, they refer to what is called a fixed action pattern (FAP).

A fixed action pattern of behavior describes an activity like that of the spinning spider. Once a stimulus has started her spinning, she will continue automatically in a step-by-step process no matter what happens. In other words, a fixed action pattern ignores feedback from the senses and makes the animal continue. For example, after a goose lays her eggs, she uses her neck to pull them together into a clump. If an egg is quickly removed, she will continue to use her neck to pull at the nonexistent egg until completion. In other words, she is oblivious to her senses, which would tell her that there is no egg to pull.

Scientists have found that fixed action patterns are begun by what they call a sign stimulus or releaser. This is a type of cue in the animal's environment that triggers what might be described as genetically pro-

These baby birds instinctively open their mouths in anticipation of food when an adult bird returns to the nest. Because of this instinct, the baby bird is able to compete for limited resources, increasing chances for survival. (Reproduced by permission of The Stock Market. Photograph by Roy Morsch.)

grammed behavior. Although this is still not fully understood, it is known that certain "releasers" trigger a reaction in the animal's central nervous system called the innate releasing mechanism (IRM). However, this scientific explanation cannot go beyond saying only that the IRM is genetically encoded.

Despite people's inability to fully understand it, they know that instinct serves primarily to help an animal survive, and that it is controlled by its genes (meaning that instinct was inherited and will also be passed on to later generations). Instinctive behavior patterns can be modified, or changed, and can even be used to trick the animal. However, for the most part, natural selection (the process of survival and reproduction of organisms best suited to their environment) has found that an automatic response in certain important situations is best suited to assure the survival of the species. Thus a certain environmental cue, or releaser, always causes an appropriate biological response, or trigger. This permits the animal to perform the "right" action immediately.

[*See also* **Genetics; Inherited Traits**]

Integumentary System

The integumentary system of an organism is its protective outer covering. All organisms have an integument or covering that separates the organism from its environment and serves several other important functions. The integument of vertebrates (animals with a backbone) is called skin. Skin can vary widely, from the impenetrable shell of an armadillo to the amazingly smooth skin of a porpoise.

Every organism has some sort of covering that holds together its body organs and fluids and makes it separate from its environment. This outer covering protects it from foreign bodies and matter and sometimes allows it to communicate with the world outside itself. In both one-celled organisms and plants, the integument is the same as its cell membrane and any secretion or coating that it produces. More complex invertebrates (animals without a backbone) have an integument that consists of a single layer of cells called an ectoderm. Only in vertebrates is the integument a many-layered, complex organ system that serves many functions.

THE INTEGUMENTARY SYSTEM OF PLANTS

The integument, or outer covering, of plants does the same thing that skin does for animals—it protects plants from injury and prevents underlying tissue from drying out. Higher plants, or those that have seeds and a vascular system (an internal system of tubing that carries fluid), have a living epidermis usually one cell thick. It may be thin, like the covering on lettuce leaves, or thick and tough like that on pine needles. The epidermis has openings called stomates that regulate temperature and water loss. It can have coatings like the wax on an apple or sensitive hairs like those on a Venus's flytrap. Overall, most plant integuments are fairly fragile compared to animal coverings.

THE INTEGUMENTARY SYSTEM OF INVERTEBRATES

The integument of more complex invertebrates is usually a single layer of cells that secrete some type of cuticle. The cuticle oozes out of the epidermis and hardens, affording it some type of protection. In crustaceans like crabs and lobsters, this functions as an external skeleton and is very hard and tough. Insects have an outer covering made up of chitin fibers that is secreted by the epidermis. Chitin forms a type of flexible, natural plastic and acts as an outer skeleton to which muscles are attached. In insects, the cuticle is a living structure and can produce sensitive hairs as well as bristles, scales, claws, or wings, depending on the species. Insects grow by shedding the cuticle and growing a new, larger one. Mollusks, like clams and snails, also form a hard, external shell.

INTEGUMENTARY SYSTEMS IN VERTEBRATES

Only in vertebrates, however, is the integumentary system considered to be a vital organ. That is, it not only provides protection for the delicate tissue underneath, but it gathers and conveys information to the organism itself about the outside environment. The skin of all vertebrates consists of two layers: the relatively thin (outer) epidermis, and the tough, inner dermis. The epidermis is several cells thick, and its outermost layer of cells is made up of dead cells composed of keratin, the protein found in hair, nails, claws, beaks, feathers, scales, and quills, among other things. Vertebrates replace this outer layer of dead cells every twenty-eight days, with the new layer identical to the old. In mammals, the inner dermis is highly developed, and is richly supplied with blood vessels, glands, and nerve endings. Besides acting as a barrier against infection and retaining the body's fluids, the dermis also has a regulatory function of letting the organism know whether to raise or lower body temperature or to move to where it is cooler or warmer. In endothermic (warm-blooded) animals, the skin plays an important role in regulating the body's temperature. In humans, the epidermis also contains a dark pigment called melanin that protects the skin from the Sun's ultraviolet radiation. It is the amount of melanin an individual's body produces that accounts for what we describe as the many different colorings of human "races."

Mammals. In mammals, the presence of hair is a distinguishing characteristic. As an outgrowth from a mammal's skin, hair grows from a pit in the dermis called a hair follicle. This pit also has a small gland that secretes an oily substance that keeps the hair oiled. For most mammals, the hair's main function is to act as insulation against the cold. Hair also serves as a sensory organ (like long whiskers) for certain night-prowling animals. Eyelashes in humans serve to make the eyes reflexively shut if they are hit by a speck of dust. There are also other glands in the dermis that keep the skin oiled and waterproof. Humans also have sweat glands in the dermis that act as a temperature control by means of evaporation, and most mammals also have dermal glands that produce odors that are thought to be a form of sexual communication.

In humans, the skin is considered the largest organ of the body. It changes considerably over time, and as it ages it becomes less elastic and more wrinkled. As with all vertebrates, human skin provides both protection from and communication with its environment. The dermis is rich with nerve fibers that can respond rapidly to changing environmental conditions, reporting its findings to the brain, which makes the necessary ad-

A cross section of the skin, part of the integumentary system. Structures used for sensing are labeled on the right. (Illustration of Hans & Cassidy. Courtesy of Gale Research.)

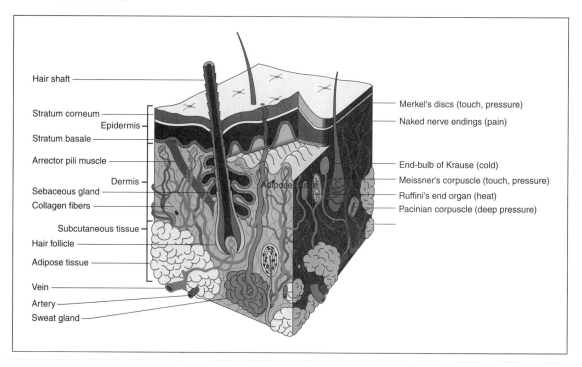

Hair shaft

Stratum corneum
Epidermis
Stratum basale

Arrector pili muscle
Dermis
Sebaceous gland
Collagen fibers

Subcutaneous tissue
Hair follicle
Adipose tissue

Vein
Artery
Sweat gland

Merkel's discs (touch, pressure)
Naked nerve endings (pain)

End-bulb of Krause (cold)
Meissner's corpuscle (touch, pressure)
Ruffini's end organ (heat)
Pacinian corpuscle (deep pressure)

Adipose tissue

justments. Besides its practical functions as barrier and regulator, the human skin also possesses the exquisite and indescribable sense of touch. Its surface is usually home to many bacteria, mostly harmless, and the skin can be subject to many diseases or injuries.

Invertebrates

An invertebrate is a multicelled animal that does not have a backbone. Within this seemingly simple grouping there is an amazing variety of complex life forms: from sponges and starfish to earthworms, clams, spiders, and butterflies. Of the roughly 1,500,000 different species of animals in the world, more than 95 percent are invertebrates. They inhabit nearly every type of environment on Earth and vary greatly in the way they live and reproduce. An invertebrate may be as soft as a jellyfish or as hard as a lobster but they have one distinction in common—they have no bony vertebral column or backbone.

In nature there is no actual dividing line that separates animals with backbones from those without one, but grouping the members of the kingdom Animalia in this manner allows biologists to sort them into very broad groupings. The animal kingdom is divided into major groups called "phyla," (singular, phylum), and of all the animal phyla identified (some say there are as many as thirty-eight), only one includes vertebrates. The rest are invertebrates. This gives some sense of how successful these "lower" animals have been in the race for survival. Invertebrates not only live almost everywhere on Earth, but range in size from an organism too small to be seen without a microscope to a giant squid measuring 60-feet (18.29 meters). Invertebrates are often considered to be pests, yet despite our best efforts to exterminate them, they seem to adapt and thrive.

SPONGES

The sponge is the simplest of all invertebrates (and therefore the simplest of all animals) and lives at the bottom of the sea. Early naturalists considered sponges to be plants since they looked like a plant and did not move. Later, as they were studied more, sponges came to be considered "zoophytes" or plant-animals. Today they are considered to be the simplest of animals and are placed in the phylum Porifera. The light brown, oddly shaped sponge we sometimes use to bathe ourselves or wash the car began its life attached to the bottom of one of Earth's seas, where it used its many holes or pores to draw in water and filter it for food. Sponges have bodies resembling a sack with an opening at the top.

They have no organs or nervous system, and usually reproduce sexually (with sperm fertilizing an egg). However, some sponges can also reproduce asexually. For example, when a sponge piece breaks off, floats away, and happens to settle in a proper place it begins to grow. Sponge "farmers" cut up living sponges and place their pieces on a rock underwater where a full-sized sponge will grow in a few years. The sponge we use in the bath is the dried (but very absorbent) skeleton of an invertebrate that was once alive. Sponges are so different from all other animals that some biologists believe that they should have their own animal subphylum.

CNIDARIANS

One step up the invertebrate ladder of complexity from sponges are the members of the phylum Cnidarian (the "C" is silent). Also called coelenterates (Latin for "hollow gut"), cnidarians include jellyfish, sea anemones, corals, hydra, and their relatives, all who live in the water. Although they are not related to sponges, cnidarians are usually listed after sponges up the invertebrate ladder because they started with the sponge's simple multicelled form and added many features not found in sponges. One of the things cnidarians added was a "hollow gut" or specialized digestive cavity, which is attached to a type of mouth (that also serves as an opening to expel waste).

A cnidarian has only one opening that serves two purposes: to take in food and expel waste. All cnidarians have armlike projections called tentacles that hang down around their mouth. When a small fish bumps into them, the tentacles react and sting or grab the fish, reeling it into its mouth. Since cnidarians are designed to be organized around a single food-gathering mouth, their body form is described as having "radial symmetry." This means that the cnidarian body has no definite right or left side but resembles spokes radiating from the hub of a wheel. Some cnidarians, like coral, are filter feeders and stay in one place, while others, like the jellyfishes, can swim around.

Cnidarians reproduce sexually, but they can also duplicate themselves asexually by budding or producing new cells that separate from the parent and become independent organisms. Cnidarians are named because of specialized nerve cells called "cnidoblasts," which makes their stingers work. Stinging tentacles are used to get food and for defense. These tentacles are not linked by a central nervous system but operate independently and almost automatically in a stimulus/response manner. This is why a dead jellyfish can continue to inflict a bad sting to someone touching its tentacle.

WORMS

Worms are more complex than sponges or cnidarians, and although the word worm refers to any animal that has a long, soft body without legs, a worm is far from a simple animal. The most important difference between worms and the other two invertebrate types is not obvious, however. Instead of having two layers of cells in their bodies, like sponges and cnidarians, worms and other "higher" animals have three layers of cells. Because of this middle layer between its external and internal layers, the worm has specialized tissues and organs that sponges and cnidarians do not have. The many different kinds of worms are gathered into three groups: flatworms, roundworms, and segmented worms. All members of the last two groups show bilateral symmetry, meaning that if they were cut down the middle, there would be two matching halves.

Flatworms. Flatworms, which belong to the phylum Platyhelminthes, move about for their food and therefore have a definite front and back end. This means that they have developed a head or at least a forward part in which nerves and senses are concentrated. Along with a nervous system, flatworms also have a separate excretory system and a reproductive system. The simplest flatworm is flat and always found in water. The most common is the planarian, which has only one opening to take in and

A millipede is a good representative of invertebrates. (Reproduced by permission of Field Mark Publications. Photograph by Robert J. Huffman.)

expel waste. A tapeworm is a parasitic flatworm that lives in the digestive system of its host where it attaches to an intestine wall and absorbs already-digested food. Flatworms reproduce sexually and asexually, and if cut in half, both parts will grow into a complete animal.

Roundworms. Roundworms have a cylindrical body, a tough outer cuticle, and are found in both water and soil. Also called Nematoda, many are parasitic. Pinworms are a well-known roundworm, as are the trichina that causes trichinosis (a disease caused by eating infected pork or meat). Roundworms reproduce sexually and have a separate digestive and circulatory system.

Segmented worms. Segmented worms or true worms belong to the phylum, Annelida which means "little rings." An annelid's body is therefore made up of segments or little rings attached together. The earthworm is a good example of a typical annelid in that its body is more complicated than that of other types of worms. Their digestive system contains organs with special jobs, and their nervous system has a distinct brain in the front end or head. Since each body segment has a set of muscles, annelids can slowly move about by changing their segment shapes. A very important annelid advance is the development of a "coelum," a lined body cavity that not only provides support but allows organs to be suspended inside the body. The coelum is found in all the more complex animals, including humans. Most annelids live in the soil, and some, like leeches, are parasitic. Reproduction in annelids is usually sexual. Although an earthworm is both male and female, it must mate with another earthworm before each can lay eggs.

MOLLUSKS

Although the next invertebrate phylum Mollusca means "soft," most of these invertebrate have a hard shell. Clams, oysters, and scallops are all mollusks as are squid, octopus, and snails. Despite the shell of some, they all are soft-bodied with some form of covering or mantle. In some, it is hard tissue and in others it is a very hard shell. Most mollusks have some sort of foot or appendage for feeding and moving about. Besides a digestive and circulatory system, they also have a well-developed nervous system, and some even have eyes. Finally, mollusks that live under water have specialized gills that take oxygen out of the water and put it into their blood. Mollusks also reproduce sexually.

ECHINODERMS

The phylum Echinoderm means "spiny skinned" and is made up of invertebrates like starfish and sand dollars that have hard outer plates.

These plates usually cover a body that has five separate parts like the spokes of a wheel. Echinoderms live in the sea and have a specialized system of canals in their bodies that connect to their many tube feet. These feet suck in water and allow the echinoderm to attach itself to something solid (like a clam shell which it can then pry open). The echinoderm has a complete digestive system although it has no excretory or respiratory system. All echinoderms reproduce sexually. They also are able to grow back any lost parts through regeneration.

ARTHROPODS

The phylum Arthropoda is considered the largest and most successful phylum in the kingdom Animalia. Arthropods live nearly everywhere on Earth. All have at least three pairs of jointed legs and a body divided into jointed segments that is covered by a hard, outer case called an exoskeleton. Arthropods have internal body systems that break down food, take in oxygen, circulate blood, and carry away wastes. Most reproduce sexually. Arthropods are so varied that there are five major types: crustaceans (lobsters, crabs); arachnids (spiders, ticks, mites); insects (bees, ants, beetles); centipedes; and millipedes.

As the largest animal group, invertebrates are an essential part of every ecosystem. Humans could not function without them since they are responsible for the decomposition of organic waste, which allows the recycling of essential chemicals. Invertebrates are also involved with the pollination of plants and are a crucial link in the food chain where herbivores (plant-eating animals) convert the energy in plants into a form useful to meat-eating animals.

[*See also* **Arachnids; Arthropods; Crustaceans; Insects; Mollusks; Protozoa**]

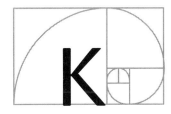

Karyotype

Karyotyping is a diagnostic tool (a way of identifying a disease or condition) used by physicians to examine the shape, number, and structure of a person's chromosomes (a coiled structure in a cell's nucleus that carries the cell's DNA) when there is a reason to suspect that a chromosomal abnormality may exist. A karyotype is made by arranging pictures of the chromosomes in matching pairs (called homologous pairs) according to their size, shape, and length. This particular technique was developed because chromosomes are very difficult to observe in a cell unless the cell is about to undergo division. When division is about to happen, the chromosomes that could be seen as only long, tangled threads suddenly begin to shorten, thicken, and condense.

Chromosomes are easily stained when they are in this form. Staining them with a dye not only makes them easier to see but also makes their distinctive bands show up well. The technique used by biochemists to make a karyotype is now fairly routine. Skin cells are often used since they divide frequently. Mitosis or cell division can be started by adding the proper stimulating chemical to the cells that are in a liquid. Once the cells begin to divide, another substance is added to fix, or freeze, the division, and the fixed cells are placed on microscope slides and dyed so that the light and dark bands on the chromosomes show up. These bands indicate the position of deoxyribonucleic acid (DNA) along the chromosomes. A photograph of the chromosomes is then taken and enlarged. Individual chromosomes are cut out and paired up or arranged by size, shape, and the length of their "arms." Biochemists use the pattern of stained bands as well as chromosome length and the position of the centromere (the strand linking two chromosomes) to identify pairs of matching chromosomes. All the chromosomes are then numbered and arranged in order, from the largest to the smallest. Today, the use of fluorescent stains, microscopes, and computers make karyotyping an easy, standard way of searching for and identifying genetic disorders.

In the hands of a trained professional, a karyotype can provide a great deal of genetic information. Some problems can be spotted with little analysis, such as Down's syndrome. This condition, which causes mental handicaps and certain facial and body characteristics, is easily seen on a karyotype. It shows up as three copies of chromosome number twenty-one instead of the normal two chromosomes. Karyotyping identifies extra or missing chromosomes as well as banding irregularities.

Doctors often recommend making a karyotype of a fetus when there is a high risk of genetic disorders. Cells are obtained from the fetus in the mother's womb using a technique called amniocentesis. In this procedure, a long thin needle is inserted through a mother's abdomen and into the fluid-filled membrane surrounding a developing fetus. Karyotyping can also be done after a baby is born to determine if a certain physical disability that it has was caused by a problem in its chromosomes.

[*See also* **Chromosomes; DNA; Genetics; Genetic Engineering; Gene Therapy**]

Kingdom

The term kingdom is one of the seven major classification groups that biologists use to identify and categorize living things. These seven groups are

Short region of DNA double helix	2 nm
"Beads-on-a-string" form of chromatin	Nucleosome 11 nm Histone
30-nm chromatin fiber of packed nucleosomes	30 nm
Section of chromosome in an extended form	300 nm
Condensed section of chromosome	700 nm
Entire duplicated chromosome	1400 nm

hierarchical or range in order of size. The kingdom group is the first and largest group. The classification scheme for all living things is: kingdom, phylum, class, order, family, genus, and species.

As the broadest of all classification groups, kingdom is made up of phyla (singular, phylum). From the time of Greek philosopher and scientist Aristotle (384–322 B.C.) to the mid-twentieth century, only two kingdoms were recognized—Animalia (animal) and Plantae (plant). However, with increasingly modern and sophisticated techniques, biologists eventually came to recognize a five kingdom approach. These additional kingdoms were necessary in order to include other forms of life that did not belong in either the plant or animal kingdom. Today, the five kingdoms are monerans, protists, fungi, plants, and animals.

Monera, the smallest kingdom (with only about 4,000 species), includes the prokaryotic bacteria (single cells that do not have a nucleus) and certain types of algae. Bacteria play an important role as decomposers, and some monerans can make their own food through photosynthesis. Organisms in the kingdom Protista are eukaryotic (their cells contain a nucleus) but are both plantlike and animal-like. Some algae and protozoans are protists. The kingdom Fungi consists of molds, yeasts, and mushrooms. These are all multicelled organisms that live by absorbing food. Although these organisms look like plants, they do not make their own food. Members of the kingdom Plantae make their own food and often grow flowers and form seeds. The kingdom Animalia includes multicelled organisms that live by taking in food. Animalia is made up of vertebrates (animals with a backbone) and invertebrates (animals without a backbone). This is the largest kingdom, containing more than 2,000,000 species.

[*See also* **Class; Classification; Family; Genus; Phylum; Species**]

Opposite: A figure of a human karyotype. Karyotyping is often a good way of detecting possible genetic disorders in unborn children. (Illustration by Hans & Cassidy. Courtesy of Gale Research.)

Lactic Acid

Lactic acid is an organic compound found in the blood and muscles of animals during extreme exercise. It also is produced in some plants as a result of the fermentation (the process of splitting complex organic compounds into simple substances) of certain carbohydrates. A buildup of lactic acid in the body is toxic and causes muscle fatigue, pain, and cramps.

Athletes who push themselves or exert their muscles and bodies beyond what is a comfortable level of exercise often experience a "burn" in their muscles. If they do not stop exercising, they sometimes will feel muscle cramps, pain, and overall fatigue or exhaustion. This is the direct result of lactate, or lactic acid, building up in the blood and muscles. This occurs naturally when a person's muscles are not given a rest and are made to keep contracting (which is how a muscle works).

During this or any kind of exercise, the body uses up energy that it gets from a process called respiration. This is not the respiration we refer to as breathing, but rather it is the chemical process of breaking food down to release the energy it contains. After we eat and our body's enzymes break down the food we have consumed into glucose (sugar), our bodies store the glucose if we do not need to use it immediately. This depends on our level of activity. A certain amount of glucose will always be needed just to maintain all of our systems and keep everything "running."

When we start to increase our body's energy demands by doing something strenuous like running or exercising, the process of aerobic respiration starts up. Our cells break down the glucose the body has stored by combining it with oxygen. Aerobic respiration releases a large amount of energy. Sometimes we exercise so vigorously or steadily that our muscles start to consume or use up the much-needed oxygen faster than we can take it in, which is why we breathe heavier and faster as we exert ourselves. When this "oxygen debt" occurs and we still do not stop exercising and consuming energy, our cells get the message that they should begin the alternative process called anaerobic respiration.

Anaerobic respiration involves the release of energy without needing to consume any oxygen (anaerobic means no oxygen). This is basically the same process that occurs in fermentation, since fermentation is the breaking down of organic materials without the consumption of oxygen. Although anaerobic respiration provides the muscles with much-needed energy, it also has a toxic or poisonous by-product called lactate or lactic acid. If lactic acid is allowed to buildup in muscles, it causes cramps and pain, makes them work less efficiently, and eventually will simply shut them down.

Lactic acid will only begin to be broken down and removed from the system once the body is able to begin normal, aerobic respiration (with oxygen). Once the body's "oxygen debt" is replenished by enough heavy breathing and an activity slowdown, the unpleasant side effects caused by lactic acid will disappear. Fitness training is thought to increase an athlete's tolerance to lactic acid buildup.

Lactic acid is also found outside the body. Not surprisingly, it is an important acidic component of fermented food products such as yogurt, buttermilk, sauerkraut, green olives, and pickles. The formation of lactic acid in these food products is the result of the activity of lactic acid bacteria. Lactic acid also has industrial uses, as it is used in a variety of processes including tanning (converting animal hide into leather) and wool dyeing.

[*See also* **Blood; Carbohydrates; Fermentation**]

An illustration of the molecular structure of lactic acid, an organic compound found in the blood and muscles of animals during extreme exercise. (Illustration by Hans & Cassidy. Courtesy of Gale Research.)

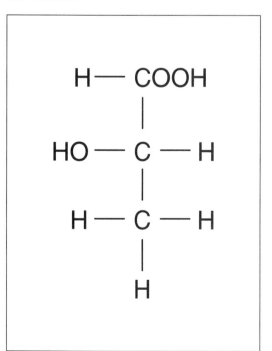

$$H \!-\! COOH$$
$$HO \!-\! C \!-\! H$$
$$H \!-\! C \!-\! H$$
$$H$$

Larva

A larva is the name of the stage between hatching and adulthood in the life cycle of some invertebrates (animals without a backbone). A sexually immature organism that lives on its own, a larva seldom resembles its final adult form and usually has entirely different life habits.

A larva, like a caterpillar, is sometimes thought to be a complete, separate, sexually mature organism that has a life of its own and produces more caterpillars. On the contrary, a caterpillar is only one stage (the larval stage) between the hatched egg of a butterfly and the adult butterfly itself. This is typical of one of the major characteristics of larvae—they seldom resemble their final adult stage.

There is no better example than that of the fat, slow-moving, hairy caterpillar that spends all its time eating, and the graceful, often beautiful butterfly that flits and darts from flower to flower. Because a butterfly is an invertebrate that undergoes metamorphosis (a total change in its body shape) as part of its development, it must pass through a larval stage (caterpillar) before it can become a sexually mature butterfly. Like the butterfly, moths also live as caterpillars before they reach their adult flying stage. Among several other invertebrates that pass through larval stages are bees, wasps, and beetles (as grubs), flies (as maggots), mosquitoes (as wrigglers), and frogs and toads (as tadpoles).

A moth larva on a leaf, which it uses as food. Larva are eating machines, since their main goal is to grow and develop as much as possible. (Reproduced by permission of Field Mark Publications. Photograph by Robert J. Huffman.)

All these types of larvae are eating machines, since their main goal is to grow and develop as much as possible. For example, caterpillars

have powerful jaws for chewing leaves. They also do not need to move quickly (nor can they), since they often simply attach themselves to their food source. While this makes them vulnerable to predators, they use different strategies to avoid being eaten. Some use camouflage and blend in perfectly with the leaf color of their favorite food. Others arm themselves with prickly hairs that irritate or with sharp spines. Some are poisonous or taste bad. Larvae also seldom eat the same thing or live in the same habitat that they will as an adult. This means that the immature organism does not compete with the adult organism.

At some point in its short life as a larva, the invertebrate will receive a hormonal signal that will trigger the beginning of its metamorphosis into an adult. According to the type of invertebrate, the larva will go through either complete or incomplete metamorphosis. Incomplete metamorphosis has three stages (egg, larva, adult). Complete metamorphosis has four stages (egg, larva, pupa, and adult). Insects like grasshoppers, dragonflies, and termites go through the shorter, three-stage or incomplete metamorphosis. For these insects, the larva are often called nymphs since they actually resemble miniature adults in their larval stage. The nymph or larva changes into an adult by molting or shedding its outer skin several time as its internal systems develop and enlarge.

Complete metamorphosis is much more dramatic since the adult that finally emerges is so drastically different from the organism it used to be. During complete metamorphosis, the larva goes through a resting stage called the pupa, during which all of these changes take place. At the beginning of the pupal stage, the larva attaches itself to something solid, sheds its skin, and forms a tough outer case around itself called a chrysalis. Usually this is a hard shell, but it sometimes can be a silken covering called a cocoon. The changes that go on during pupation consist mostly of breaking down cells and developing new cells. Eventually the adult insect is formed inside the pupa and it escapes from its casing by breaking it open. What emerges is the former larva now transformed into an adult.

About 90 percent of all insects undergo complete metamorphosis. Certain other invertebrates, like sponges, have larval stages as a way of dispersing their offspring. For example, after sexual fertilization (union of male and female sex cells), many sponges develop thousands of tiny, free-swimming larvae that are released to be carried away by currents and to finally settle and attach themselves to the ocean bottom.

[*See also* **Life Cycle; Metamorphosis**]

Leaf

A leaf is the main energy-capturing and food-producing organ of most plants. Nearly every plant on Earth owes its continued existence to its leaves, which collect energy from sunlight and convert it into food through a process known as photosynthesis. Leaves may vary widely in size and shape, but all are designed primarily to capture as much light as possible without drying out.

PARTS OF A LEAF

Leaves are attached to and supported by the plant stem, which provides the leaves with water and inorganic nutrients from the soil. Most leaves have two main parts: the blade and the petiole. The blade, or lamina, is the broad, flattened surface of the leaf that absorbs radiant energy from the sun. The blade is attached to the stem by a stalk called a petiole that also supports it. The blade is made up of two layers of cells—a tight, outer layer of cells called the epidermis, and a thicker, inner layer of mesophyll cells. The epidermis is covered by a waxy coating called a cuticle that helps cut down water loss from the leaf. It is in the inner mesophyll cells where photosynthesis is carried out.

The petiole not only joins the leaf to the stem, but contains tiny tubes that connect with veins inside the blade. Besides strengthening the blade, the veins' main purpose is to act as pipelines and transport water and food to and from its cells. A large vein called the midrib usually runs along the center of the leaf and smaller branching veins run out to its edges.

Edges of leaves can differ greatly, and while narrow leaves like grass have smooth edges, many broadleaf blades have jagged points called teeth at their edges. In some plants, these teeth act as valves and release excess water, while in others they function as tiny glands producing a liquid that repels insects. Leaves also contain a stoma (plural, stomata), which is critically important to the leaves' operations. Because the wax coating of its blade is not porous, leaves have developed special openings that allow gases (carbon dioxide and oxygen) to be exchanged and water to be released. A stoma is similar to a tiny slit that opens or closes by the action of two guard cells on either side. These cells can change shape and make the stoma open or close. This is essential during photosynthesis when the plant must take in carbon dioxide (and give off oxygen as a by-product). The stomata also enables a plant to regulate how much water it loses. When the stomata are open, they allow water to escape into the atmosphere. To minimize water loss, stomata tend to close at night when photosynthesis is not occurring, and open during the day when rapid gas exchange is necessary. During unusually dry conditions the stomata may close to prevent wilting, and photosynthesis is reduced.

LEAVES ARE ESSENTIAL TO PLANTS

Leaves are like a food factory for a plant since they begin with raw materials and process them internally to produce glucose, which the plant uses for growth, development, and reproduction. The plant itself can also become food for primary consumers that eat parts of the plant and obtain the energy the plant has stored. Within the leaf's internal structures called chloroplasts are the plant's main light-absorbing compound called chlorophyll. It is this pigment that gives plants their typically green color. Photosynthesis begins as the leaf lets carbon dioxide in through its stomata and obtains water from its roots through its veins. When sunlight strikes the chlorophyll in the chloroplasts, light energy splits the water into hy-

A close-up photograph of a tulip tree leaf showing the veins, which transport water and food to and from the leaf's cells. (Reproduced by permission of Field Mark Publications. Photograph by Robert J. Huffman.)

drogen and oxygen. Hydrogen combines with carbon dioxide to make the simple sugar glucose, and oxygen is released through the stomata as a by-product. All of this occurs within the leaf at the cellular level.

HOW LEAVES DEVELOP

At the very beginning of their existence, leaves are contained in embryo form within the seed and are called a cotyledon. Once the seed germinates, or sprouts, the cotyledon emerges and eventually becomes the first true leaf. As the plant matures, more leaves are formed from buds that formed on the stem. Once the bud begins to unfold and open, the leaf begins its growth period and reaches full size anywhere from one to several weeks. The mature leaf turns a deep green and begins to make food for itself and the rest of the plant. Although a leaf contains other colors, they are masked by the chlorophyll (a green pigment). As autumn approaches, however, the plant releases a hormone and the chlorophyll starts to break down and eventually disappears from the leaves, allowing the remaining colors of yellow, orange, or red to finally be seen. Once the chlorophyll breaks down, the leaf no longer makes food, its veins become plugged, and it soon withers and dies. A layer of cells grow across the base of its petiole, shutting it off from the stem, and the leaf soon dries, twists in the wind, and breaks off. When a tree's dead leaves fall to the ground, they take away some of the waste products the tree produced. They also eventually become food for bacteria and decay on the ground, adding essential humus or organic matter to the soil and offering new nourishment for other plants to use.

Leaves are vital to life on Earth. Since they are the actual site where photosynthesis occurs, they are the first link in the food chain (the series of stages energy goes through in the form of food), providing food to animals. They are not only the factories of the primary producers (plants), but they help make the air breathable for animals. Without the oxygen that plants give off during photosynthesis, Earth's supply of breathable oxygen might be eventually used up. People also use leaves for many products, from tea and herbs, to lettuce and spinach, to drugs like digitalis and tobacco.

[*See also* **Photosynthesis; Plant Anatomy; Plants**]

Life Cycle

The term life cycle describes the series of predictable changes that an organism goes through until it is mature enough to reproduce. Knowledge of the major stages or changes that all species undergo during their lives is essential to the study of the life sciences. Studying an organism from birth to sexual maturity is an ideal way to learn what is most important and essential to its life and continuance.

For some species, a complete life cycle is only fifteen days, while for others it can be decades. However, during the normal life cycle of every organism, growth and reproduction always take place. Between birth and sexual maturity, some species go through a long sequence of basic changes over time while others appear to make a direct trip. For example, although mammals are relatively complex animals, their life cycle is fairly straightforward. Mammals begin to develop from a fertilized egg and once born, they simply continue to develop or grow. There is certainly much variation between mammals, since a human baby takes about eighteen months to learn to walk, while a horse will stand up almost immediately at birth and romp in a day. Childhood for mammals also varies in length. Humans enter puberty (the stage at which they begin to mature sexually) in their early teens, while a dog may be ready to have a puppy before it is a year old. Despite these differences, the life cycle basics are nearly the same for all of the higher animals (birth, growth and sexual maturation, fertilization, birth).

INCOMPLETE METAMORPHOSIS

While some lower organisms have simpler life cycles, there are many animals and plants with life cycles that are not so straightforward. Some animals go through complex life cycles in which they physically become an entirely different type of individual. In other cases, a period of asexual reproduction (without the union of sperm and egg) is followed by a period of sexual reproduction. For example, a grasshopper has a three-stage life cycle called incomplete metamorphosis. After an adult female grasshopper lays an egg and buries it, the egg develops and eventually hatches. What emerges from the shell is called a nymph. At this stage in its life, the nymph may look like a miniature adult but it has no wings and no working reproductive organs. As the nymph grows, it periodically sheds its outer skin or molts, and with every molt it becomes more of an adult. When it sheds its skin for the fifth and final time, it has become an adult grasshopper and is ready to mate and reproduce.

COMPLETE METAMORPHOSIS

Other insects, like a moth or butterfly, go through a much more complicated process called complete metamorphosis. After an adult female moth lays its eggs and they develop and mature, what hatches looks like

Opposite: A labeled diagram of a butterfly's life cycle, which is a good example of complete metamorphosis. (Illustration by Hans & Cassidy. Courtesy of Gale Research.)

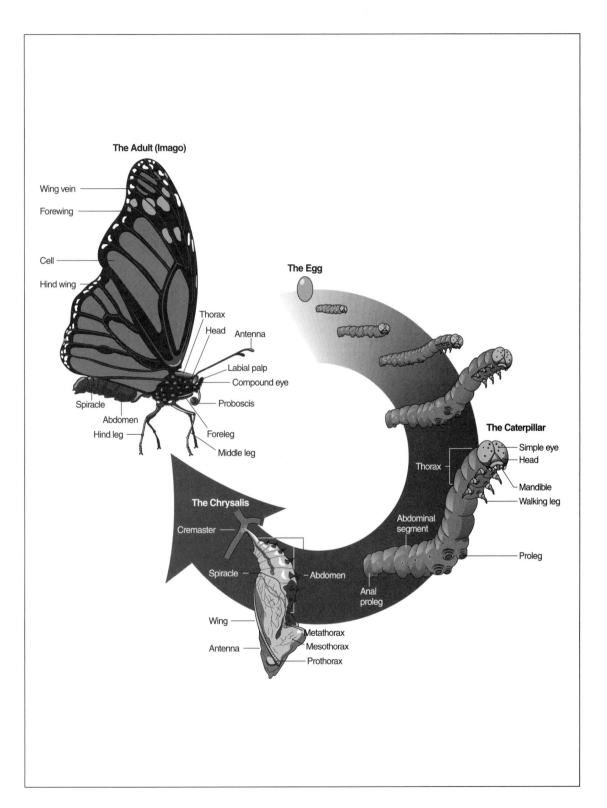

The Adult (Imago)

Wing vein

Forewing

Cell

Hind wing

Thorax

Head

Antenna

Labial palp

Compound eye

Spiracle

Proboscis

Abdomen

Hind leg

Foreleg

Middle leg

The Egg

The Caterpillar

Simple eye

Head

Mandible

Thorax

Walking leg

Abdominal
segment

Proleg

Anal
proleg

The Chrysalis

Cremaster

Spiracle

Abdomen

Wing

Metathorax

Antenna

Mesothorax

Prothorax

a worm and is called a larva. A larva is the caterpillar stage in a moth's life cycle. As a caterpillar, the larva is nothing more than an eating machine, and its body is built to help it consume as much food as possible. It has a long body with three pairs of true legs. It also has a large head with strong jaws that allow it to feed on plants. Many caterpillars have some form of camouflage or coloring that allows them to blend in with the plants they eat. Others may have bright warning colors and irritating hairs that keep predators away. After a series of molts, or outer skin shedding, the caterpillar produces an outer covering around itself called a cocoon and attaches itself and the cocoon to a branch. Inside these coverings, most of the larva's cells are broken down and begin to reform as a pupa. As the pupa develops inside, it is reformed and transformed into an adult insect, and a moth or butterfly emerges. As an adult, the insect is soon ready to reproduce and its life cycle is complete.

ALTERNATION OF GENERATIONS

Discovering the details of an organism's life cycle can sometimes be essential to understanding its true nature. For centuries, no one was able to discover how ferns reproduced. It was long thought that since a fern was a green plant, it had to produce seeds (and therefore reproduce sexually with male and female sex cells). Yet finding a fern's seeds proved impossible. Botanists (people who specialize in the study of plants) were only able to solve this problem by closely studying a fern's life cycle. It was finally discovered that ferns, as well as other plants like mosses, reproduce by spores and not seeds. Also, it was found that a fern has a sexual stage that alternates with an asexual stage that produces spores. This process of going through two different plant forms in one life cycle is called the alternation of generations.

A fern's life cycle begins when a mature fern plant produces spores inside little cases, which are attached to the underside of the fronds (leaves). Called sori (singular, sorus), these dark brown dots are sometimes mistaken for bugs or disease spots. When the spores mature, their cases split open and the tiny, light spores are sometimes carried great distances from the parent plant by the wind. When it lands in an inviting place, the spore develops into a tiny green plant called a gametophyte and produces sperm and egg. This is the sexual stage of the fern. When sperm and egg unite during rains or with dew, a fertilized egg forms and the asexual stage of its life begins. The egg develops into a new individual spore-producing fern plant, which will begin the cycle all over again.

[*See also* **Larva; Metamorphosis**]

Light

Light is energy from the Sun that we can see. Light is essential to all life on Earth, as it is the source of food, air, and warmth. Visible light is actually made up of a spectrum of colors.

Although all life on Earth depends on light, it is easy to take it for granted. However, a world without light is almost impossible to imagine. Without light from the Sun, people's eyes would not work. Also, plants would not make their own food, which feeds other animals. Nor would plants give off any oxygen as a by-product of making food, and there would be no breathable air. Without light, there would be no warmth, and Earth would be as cold as the deepest part of outer space. On Earth, therefore, light means livable conditions and life itself.

All light is really energy that travels through space from the Sun. Sunlight is a form of energy called electromagnetic energy, or electromagnetic radiation. Physicists (a person specializing in the study of matter and energy and the interactions between the two) have long known that there are many kinds of this radiant energy that streams from the Sun in waves. The visible light from the Sun is only one type of radiant energy. The other types of radiant energy are known as gamma rays, x rays, ultraviolet, infrared, microwaves, and radio waves. The entire range of this energy, including visible light, is called the electromagnetic spectrum.

Each of these forms of radiant energy travels in waves and each has its own wavelength. A wavelength is the distance from one wave peak to the next. Gamma rays have the shortest length, while at the opposite end of the spectrum are radio waves, which have the longest length. Visible light is somewhere in between these two and is the part of the electromagnetic spectrum to which the human eye is sensitive. Although this light appears white to the average person (some may say it is clear), it is really made up of another spectrum, a spectrum of colors. Thus the visible spectrum actually includes all the colors of the rainbow.

Finally, when this light streaming from the Sun encounters matter, the light is either reflected, absorbed, or transmitted to someplace else. What determines this is the color, or pigment, of the matter the light meets. Different pigments absorb different parts of visible light. The color of a pigment is determined by the type of light that it reflects or transmits. Thus, green pigment looks green to human eyes because it transmits and reflects green light. It also absorbs red and blue light, which we do not see. A black pigment absorbs all visible light, while a white pigment reflects all colors of visible light.

One of the great discoveries of science is the first law of thermodynamics. It is also called the law of conservation of energy. It states that energy can be neither created nor destroyed, although it can be changed from one form to another. This changing of the form of energy is what enables light to play such an essential role in the maintenance of life on Earth. Light energy from the Sun can be transformed into heat energy when it is absorbed by Earth. Even more important is the change from light energy to chemical energy. This occurs during photosynthesis.

Remarkably, plants absorb less than 1 percent of the sunlight that reaches Earth. This is enough, however, to allow every plant on Earth to grow and make food through the process of photosynthesis. This chemical process begins with sunlight. It carries out a chain of chemical reactions that produces not only food for the plant but oxygen for the atmosphere. When humans breathe the air and eat their food (animal or vegetable), they are incorporating the energy from the Sun (light) into their own beings. Light is therefore truly the source of all life.

[*See also* **Photosynthesis**]

Lipids

Lipids are a group of organic compounds that include fats, oils, and waxes. Lipids are important because they are a concentrated source of energy. They also serve as an important building material for cells and have many industrial and commercial applications.

Lipids are organic or natural substances that are produced by animals and plants. Common lipids include butter, vegetable oil, and beeswax. Lipids are not soluble in water, meaning that they cannot be dissolved in it. In fact, lipids repel water. Fats and oils are both lipids, yet they are different. Fats are usually solid or semisolid at room temperature (like butter), while oils are liquid at room temperature. Lipids are classified as saturated or unsaturated depending on their chemical structure (the type of bonds between the atoms). It is these bonds that make fats solid and oils liquid at room temperature.

Animals and plants store fats in their cells to use as an energy reserve. Plants usually store lipids in their seeds, while animals store them in cells of their skin. When needed, both can convert them back into fatty acids (which are made up of carbon, hydrogen and oxygen and are therefore a basic energy source). Many mammals use this layer of fatty deposits below their skin to keep warm in cold weather. Body fat is an insulator against low temperatures and internal heat loss. It is also an excellent shock absorber. Animal fats are rich in saturated fatty acids, while plant oils are rich in unsaturated fatty acids.

Lipids are an important part of a healthy human diet and are needed for normal growth, blood clotting, and healthy skin. They also are essential to the proper hormonal functioning of many animals. People also have found many practical uses for lipids, and use them in the production of many industrial products such as cosmetics, cleaners, and lubricants. Much of the soap, detergents, and cosmetics we use are made from purified animal and plant sources.

Besides fats and oils, lipids also include waxes. Waxes are soft, slippery substances that are similar in their chemical structure to fats and oils. Waxes usually resist attack by other chemicals and are produced by plants and animals. Plants use waxy lipids to coat their leaves and fruit and to prevent moisture loss. Animal skin is covered with a waxy lipid, and lamb's wool is protected by a very soft wax called lanolin. The wax made

A transmission electron micrograph of lipid droplets in a rat fat cell. (©Photographer, Science Source/Photo Researchers.)

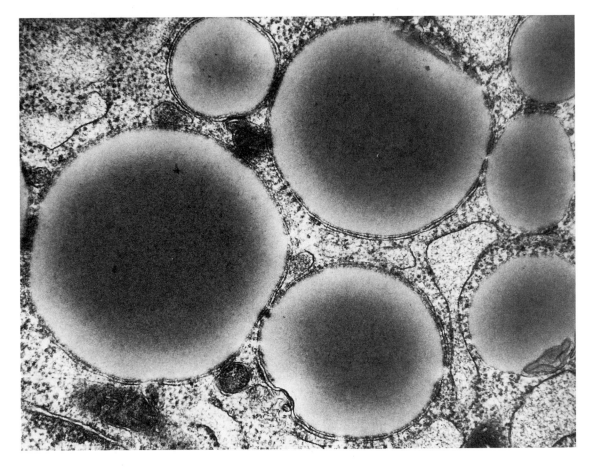

by bees is also considered to be a lipid. While lipids are generally considered a type of biological fuel since they can be converted into energy, wax can be digested by very few animals.

Although some portion of a healthy diet should include lipids, diets high in animal fats are known to cause serious health problems, such as arteriosclerosis (the abnormal thickening and hardening of the arterial walls), heart disease, and cancer. Generally, animal foods are rich in both saturated fats and cholesterol, while plant foods contain unsaturated fats and no cholesterol. Cholesterol is a kind of lipid called a steroid, and although it is essential to the body, too much cholesterol can accumulate in the arteries and cause heart disease. For this reason, it is important for humans to maintain a diet low in saturated fat and cholesterol.

[*See also* **Metabolism; Nutrition**]

Lymphatic System

The lymphatic system is a network of vessels and channels that branch throughout the body and bathe cells in a fluid called lymph while also filtering out foreign material from the blood. As the body's second circulatory system, the lymphatic system is the body's transport mechanism for necessary fluid (lymph) and also makes up part of the body's defense system by carrying lymphocytes (specialized white blood cells) that fight infection.

As blood circulates throughout the human body, a tiny fraction of its plasma (the fluid part of the blood) leaks out and collects around body cells and tissues. The lymphatic system exists partly to prevent this fluid, called lymph, from building up in the tissues of the body. It does this by a system of tubes with thin walls that absorb the excess fluid and move it slowly through the body. The lymphatic system does not have a pump like the heart to move its fluid, but instead uses the normal movements of the body to slowly push the fluid in a one-way direction. Lymph vessels do more than just transport excess fluid however, for they also pick up fat from the digestive tract and transport it to the blood where it will be used for energy. It also plays an important role in the movement of white blood cells in the body's immune response.

The complete lymphatic system includes lymph nodes and lymph ducts as well as the thymus and the spleen. The system branches to nearly all parts of the body, excluding the brain or spinal cord, draining and filtering lymph and returning it back into the bloodstream. Lymph is a colorless fluid that, although it is derived from blood, contains no red blood cells. As it seeps out of the capillaries (hairlike blood vessels) and circulates through lymph vessels, lymph is filtered as it passes through lymph nodes that are clustered throughout the body. These lymph nodes are round structures that are about the size of a pea. They act as filters, catching bacteria, toxins, dead cells and other particles, and destroying them. The main groups of lymph nodes are located in the neck, armpit, chest, abdomen, and groin. These nodes often become swollen and painful as they produce and supply extra white blood cells to fight an infection. When this happens, we experience what are called "swollen glands."

The thymus, located behind the breastbone and between the lungs, is also part of the lymphatic system since it secretes a hormone that tells the body's bone marrow (tissue that fills the bone cavities) to produce specialized white blood cells called lymphocytes. It is these cells that attack and destroy invading microorganisms as well as create antibodies (a specific protein targeted to kill a specific invader). The lymphatic system is the main highway for the patrolling lymphocytes that are always on the lookout for foreign bodies. Another part of the lymphatic system, the spleen, filters lymph and removes waste and other materials (like old red blood cells). In humans, the spleen is located below and behind the left side of the stomach. Altogether, the lymphatic system can be thought of as the body's fluid drainage network that filters and destroys foreign particles through the transportation of infection-fighting cells.

[*See also* **Antibody and Antigen; Immune System**]

Lysosomes

Lysosomes are small, round bodies containing digestive enzymes that break down large food molecules into smaller ones. They are found in the cytoplasm, or jelly-like fluid, of all eukaryotic cells (cells with a distinct nucleus). Lysosomes are the main site where digestion takes place inside a cell.

As organelles or specialized, membrane-bound structures inside a cell that have a certain job to do, lysosomes contain very powerful enzymes called "hydrolases" that are capable of breaking down many different types of substances. These enzymes work on food molecules such as proteins, carbohydrates, and fats and quickly break them down into smaller particles that can be easily used by the cell. The powerful enzymes in lysosomes are also sometimes put to another use by the cell when it needs to rid itself of a damaged or defective organelle. In such a case, the lysosomes attack an organelle and quickly break it down and destroy it. At other times, a cell may use lysosomes to actually destroy itself. This process is known as "autolysis" (*auto* means "self" and *lysis* means "destruction"). This usually happens for a very good reason, as in metamorphosis when an animal has to entirely reshape its tissues (as when a caterpillar changes into a butterfly). While biologists know that lysosomes are used by the cell to digest the food it takes in, they do not yet fully understand how the lysosome membrane itself avoids being broken down by the enzymes it carries. This is especially puzzling since the membrane is made of the same compounds that the enzymes easily destroy.

[*See also* **Cells; Enzymes**]

Malnutrition

Malnutrition is the physical state of overall poor health. It can result from a lack of enough food to eat or from eating the wrong foods. Malnutrition is most common in developing countries where people do not get enough to eat or are able to eat only one type of food.

If nutrition means eating enough of the right kinds of food to stay alive and healthy, then malnutrition is literally "bad" nutrition. Malnutrition affects a large part of the world's population. Even in developed countries, many children suffer from some forms of diet deficiency similar to malnutrition. Since animals cannot photosynthesize, or make their own food, as plants do, animals must get all their nutrients (the substances necessary for life) from their food. A balanced diet would contain all the basic types of nutrients that an animal needs. These include proteins, fats, carbohydrates, vitamins, and minerals. Malnutrition can be the result of a lack of many of these vital nutrients or of one particular nutrient.

Few people, if any, choose to be malnourished. This unhealthy condition can have one or more causes. External circumstances, such as war or a crop failure, can make it impossible to get enough food to eat. Very often, essential foods can become scarce or completely unavailable. Poor eating habits can also lead to a person eating only one kind of food and excluding many others. Finally, a physical condition can prevent or impair the proper digestion and absorption of food. For example, people in developing countries who drink contaminated water often get prolonged cases of diarrhea. They then lose essential nutrients from their bowels.

PROTEIN-ENERGY MALNUTRITION

Whatever the external reason, probably the most serious form of malnutrition is called protein-energy malnutrition. The terribly sad pictures of very young children with large, swollen bellies are examples of protein deficiency. Their huge bellies, made so large by a fluid imbalance, seem to deny the fact that they are in fact starving to death. Proteins, which are made of amino acids, are essential to human life since they are needed for cell growth and repair. Of the twenty amino acids needed by the body, ten can only be obtained through diet. Without a certain amount of these essential amino acids, a person's body systems will stop working normally, resulting in sickness and even death.

In many very poor countries, people often eat very little protein and live mainly on a diet very high in carbohydrates. This is understandable since carbohydrates, like corn and other grains, are fairly cheap to grow and process. The protein that is found in meat, fish, poultry, and milk is naturally more expensive and in shorter supply. Lacking proteins, the malnourished person's body also cannot make the enzymes (proteins that control the rate of chemical changes) necessary for all of the chemical reactions required during digestion.

When a person first begins to be severely deprived of food, called starvation, the body starts a process of using up its stores. First to be used are the carbohydrates, since the body stores very little of these. After two days or so, the carbohydrates are gone and the body turns to any stored fat it may have. Not until it has turned all of the fat it has into energy, and only after it has no more, will the body begin to turn to its own protein sources. Some describe fat as a protein protector. However, with all fat gone, the body must begin to use protein as a source of energy.

Proteins are essential for enzymes and hormones (chemical mes-

Two malnourished Sudanese children. Their huge bellies are the result of a fluid imbalance due to protein deficiency. (Reproduced by permission Reuters/Corbis-Bettmann.)

sengers) and serve as the body's building blocks. Using up this precious source without replacing it can only result in the body's cells not doing their jobs. The body, in effect, starts to consume itself. In children, this results in slowed mental and physical growth. Since their immune systems (a collection of cells and tissues that protect the body against disease-causing organisms) have shut down, these children cannot fight infections and easily contract diseases. Besides their swollen bellies, their hair often falls out and their muscles waste away. Needless to say, they are listless and have little energy before they finally die.

Adults who are malnourished show the same mental dullness as children, and they also easily contract diseases. Mothers give birth to severely underweight babies and cannot produce enough breast milk to feed them. The lack of a certain vitamin or mineral can also result in a particular deficiency disease, but this is more easily corrected than is a severe case of general malnutrition. As the world population continues to grow at an alarming rate, it appears that malnutrition may continue be a constant problem for developing countries.

[*See also* **Blood; Nutrition**]

Mammalogy

Mammalogy is the branch of zoology that deals with mammals. Major subject areas in mammalogy include anatomy (structure), physiology (function), behavior, ecology, evolution, and classification. Humans are mammals and belong to the class Mammalia which is one of the most diverse groupings of animals.

Mammals are among the smartest, fastest, and largest animals on Earth. The cheetah is the fastest mammal; the blue whale is the largest; and human beings are the most intelligent. There are 18 orders of mammals, containing about 4,000 living species. Aside from a couple of exceptions, all mammals have certain things in common. They are warm-blooded (they maintain a constant internal temperature despite their environment), have hair on their bodies, give birth to live young, and feed their newborns with milk from their mammary glands. It is because of these milk glands that mammals got their name. Beginning with the Greek philosopher Aristotle (384–322 B.C.), the Greeks were the first to systematically study, categorize, and write about mammals. In fact, it was Aristotle who recognized that both whales and dolphins were really mammals and had more in common with land-based animals than they did with fish.

Mammalogy allows us to understand how and where a mammal lives, what are its habits and behavior, and how it reproduces. Aside from the common mammal features already mentioned, there are other characteristics of mammals that deserve mention. All mammals have a basic structure. They all have skulls that house a brain, and they all have seven vertebrae (bony segments) in their necks, whether their neck is as long as a giraffe's or as short as a dormouse. Most have four limbs that end in five digits (finger-like projections). Their teeth are adapted to their feeding habits, and most carry their unborn inside their bodies until birth. Mammals have highly developed senses, although not all are so sharp in every mammal. Some rely more on keen vision, while others depend most on their sense of smell. Most mammals are herbivorous (feeding on plants), while fewer are carnivorous (feeding on meat or other animals). Mammals are found in almost all habitats, and many actually build some sort of shelter or dwelling for themselves. For example, beavers build underwater lodges, gorillas make beds of palms, and prairie dogs have underground tunnels. Some mammals hibernate, or enter a sleeplike state, during winter, while others migrate or travel some distance to avoid winter weather.

While many mammals live alone as adults, many also live in groups of different sizes. For example, humans live in families typically consisting of a male, female, and their offspring while beavers live in family groups. Monkeys live in larger groupings called bands, while sheep live in larger groups called herds. The largest grouping of mammals are called colonies, and this is how bats live. As social animals, mammals need to communicate, and they do this in many different ways. Some communicate by a scent and give off a certain smell when they are in heat (and are ready to mate). Many use visual signals. For instance, while a gorilla may make certain facial gestures, a wolf will assume a certain body stance. Most mammals, however, use sounds to communicate, from a coyote's howl to a beaver's tail-slapping.

At the beginning of the twenty-first century, it is safe to say that nearly all of the world's mammal species are known to science. Yet the same cannot be said about the biology of every species. Twenty-first century techniques and technology will give mammalogists (people who study mammals) the ability to study free-living (animals in the wild) animals by the use of data obtained from tiny radio transmitters placed on the animals. These transmitters will also allow scientists to learn more about the genetics of mammals.

[*See also* **Mammals**]

Mammals

A mammal is a warm-blooded vertebrate (an animal with a backbone) animal with some hair that feeds milk to its young. Mammals are the most diverse as well as the most successful vertebrate, and can be found living in nearly every habitat on Earth. Mammals are also the most advanced or intelligent animal and have become the dominant form of life on Earth. Humans are mammals.

The animals that make up the amazingly diverse class known as Mammalia range from the largest animal that ever lived, the blue whale that can reach 100 feet (30.48 meters) and 150 tons (136.05 metric tons) to the hog-nosed bat of Thailand that is the size of a bumblebee. Mammalia includes human beings, whose intelligence shaped the world as it is today, and the grotesque and slow-moving sloth that hangs upside-down from a tree most of its life. Rats are mammals, and so are dolphins, monkeys, and giraffes.

COMMON TRAITS OF MAMMALS

Despite the extreme differences among species of mammals, all have several traits in common.

Although they act more like fish, whales are considered mammals. The blue whale pictured here is one of the largest animals to ever live.

Mammary Glands. First, mammals get their name from the character-istic "mammary" glands that females use to feed their young. After a preg-nant female gives birth, her mammary glands secrete milk that the young drink by sucking. This liquid substance provides all the nutrients that a young mammal requires to grow and develop. Some mammal species suckle their young for only a few days, while others, like the elephant and humans, may nurse them for more than a year. A mammal also usu-ally gives birth to live young. Except for only three species out of ap-proximately 4,300, the young of every mammal develop inside the body of the mother who delivers or gives birth to live young instead of laying eggs. This makes mammals placental animals (named after the organ called a placenta which nourishes the developing embryo while it is in-side its mother's uterus). Another characteristic related to their offspring is that mammals usually care for their young, even after the young cease nursing on mother's milk. Mammal parents both protect the young from enemies and teach them necessary survival skills.

Hair and Teeth. Interestingly, body hair is a mammalian trait, and all mammals have at least some hair or fur on their bodies at some stage in their development. Even smooth-skinned whales and dolphins have hair at birth. Mammal hair and fur is made of long thin strands of protein called keratin. Many mammals have long, stiff hairs around their head or mouth that act as feelers and allow them to get around in the dark. The keratin in some mammals has adapted into a protective device, such as the quills of a porcupine. Hair insulates mammals in much the same way that feathers keep birds warm. This hair insulates and helps keep body heat from escaping. The fact that mammals can generate their own heat (as long as they have enough food to eat) means that they are endother-mic or warm-blooded. This is a distinct advantage because it means that mammals can maintain their own internal body heat despite living in a cold climate. Unlike a cold-blooded animal whose body temperature rises and falls with that of its environment (and as a result, becomes lethargic in cold weather), mammals are always ready to spring into action. This is a necessary trait if one is either the hunter or the hunted. Most mam-mals are also able to cool their bodies in hot weather because they have sweat glands on their skin that produce moisture. This moisture then evap-orates and cools the mammals' bodies.

Mammal teeth are also specially adapted to their feeding habits. Mammal carnivores (meat-eaters) have distinctive canine teeth that al-low them to catch, hold, kill, and eat their prey. Herbivores (plant-eaters), however, have flat, grinding teeth for breaking down the tough cell walls of plants. Beavers and squirrels have teeth that continue to grow, and an elephant's tusks are really its incisor teeth (what humans call their front teeth).

Senses. Mammals also have highly developed senses, although few species have acute capabilities in all five senses (sight, smell, taste, touch, hearing). Most have an excellent sense of smell since this sense is not only necessary in locating food that may be out of sight, but it is crucially important in alerting an animal to danger. Mammals also usually have an excellent sense of hearing and usually, the larger a mammal's ears, the more it relies on that sense for its survival. For example, a rabbit's primary defense is its quickness and, therefore, its large, long ears serve as an excellent early-warning system.

Skeletal System. As vertebrate animals, all mammals also share a basic skeletal structure. They all have a bony skull that houses a brain and key sense organs (such as eyes, ears, and nose). All have a backbone or vertebral column consisting of individual vertebrae. Except for sloths and manatees, all mammals have exactly seven vertebrae in their necks. Thus a giraffe has seven very large ones, and a mouse also has seven very small ones. Most mammals have four limbs, each of which usually ends with five digits (fingers and toes in humans). Mammals also have a distinctive, four-chambered heart (as birds do). This type of heart keeps oxygen-rich blood separate from the oxygen-deficient blood by having each flow in and out of the heart in its own system. Because of this system, the mammal heart can quickly deliver large amounts of high-energy oxygen on demand.

Behavior. Unlike invertebrates (animals without a backbone), mammals exhibit what is usually called complex behavior. For example, mammals act on instinct, a pattern of behavior that is inborn rather than learned. When a newborn human immediately seeks its mother's breast to suck or when a bear hibernates for the winter, these mammals are following instincts (or natural drive to do something). Mammals have larger brains than other animals and can be said to learn in some ways. If a mammal changes its behavior because of repeated experiences, then mammals can learn. When a mammal avoids a situation that was dangerous in the past and seeks another alternative that was previously beneficial, the mammal also exhibits a type of learning. Such behavior leads to the statement that mammals, and not just humans, are the smartest animals on Earth.

The Brain. Mammals have a large brain with a well-developed cerebral cortex. The cerebral cortex is the part of the brain involved in memory, sensory perception, and learning. Mammals also often live in social groups of different sizes. The smallest social group includes a male, a female, and young. Others live in larger groups called bands, and others in still-larger groups called colonies. The largest group of mammals would be composed of thousands and is called a herd. In bands of mammals, like monkeys, there is always a ranking of individuals with the most dominant member acting as a sort of boss or leader.

Communication. Mammals also establish bonds between one another. Mammals interact with one another by communicating in many different ways. Many use smells to tell friend from stranger or to mark their territory. Visual signals are also used, as when a rabbit flashes a patch of white hair under its tail (danger), or when a dog bares its teeth and lowers its tail. Sound is the most obvious way to communicate, and cries or whistles of danger are different noises than those of mating calls. Altogether, the behavior of mammals can be very complex.

Development of Young. Among the 4,300 species of mammals, all can be placed in one of three groups based on how their young develop: monotremes, marsupials, and placentals. There are only two species of monotremes, which are by far the strangest type of mammals, since they lay eggs. The duck-billed platypus and the spiny anteater lay leathery eggs instead of giving birth to live young. However, they have mammary glands and nurse their young. A marsupial is a mammal whose young complete their development inside the mother's pouch. Kangaroos and koalas, as well as wallabies and opossums, carry their young inside their bodies for a short period, after which they give birth to a tiny, barely developed marsupial that crawls into its mother's pouch where it nurses and grows. All other mammals are placental, meaning that they are nourished inside the female's body until birth. Some mammals, like the horse, are able to walk within minutes of birth, while others, such as a human infant, are helpless and require years of care.

The diversity of mammals can be amazing. For example, bats are the only mammals that can fly, and rodents are the largest group of mammals. Whales, porpoises, and dolphins are aquatic mammals (their fins are limbs) that breathe air, and elephants are the largest land mammals. This being the case, mammals are an interesting group of animals to study and learn from.

[*See also* **Mammalogy**]

Meiosis

Meiosis (may-OH-sis) is a specialized form of cell division that takes place only in the reproductive cells. The goal of meiosis is to produce sex cells (sperm and egg) that have only one set of twenty-three chromosomes. When a sex cell unites with another sex cell, the zygote (fertilized egg) will have the proper total of forty-six chromosomes.

It is important to distinguish meiosis from mitosis (my-TOH-sis). Although both are a form of cell division, mitosis produces two identical cells, while meiosis produces four different cells. Mitosis makes new, identical cells so that an organism will be able to replace damaged and dead cells and be able to grow. Nearly all the cell division in an organism can be described as mitosis. Meiosis only occurs in an organism's sex cells and is structured so that it deliberately produces different rather than identical cells. Without meiosis producing differing cells, there would be no variation in offspring, who also would have twice the number of chromosomes than they should have.

If mitosis occurred in reproductive cells the way it does in all other cells of the body, the new cell produced would have twice the number of chromosomes that it should have. For example, the exact amount of chromosomes needed to be human is forty-six. Without meiosis cutting the number of chromosomes in half, that number would be ninety-two chromosomes after fertilization has taken place. Meiosis splits in half the number of chromosomes in sperm and egg cells, so when the cells unite, the zygote will get half the number of chromosomes from each parent.

Besides halving the number of chromosomes, meiosis also performs another very important function. It allows genetic material to be "shuffled" as the chromosomes cross over each other and swap genes before the cell divides. This is a random exchange of genetic material that guarantees that an entirely new individual will be produced after fertilization. Because of this shuffle of genetic instructions, each reproductive cell is given its own unique set of instructions. This assures that no two sperm or egg cells have the same exact combination of genes. This also partly explains why brothers and sisters (except identical twins) of the same parents have different characteristics.

Eventually, when two of these unique sex cells are joined as sperm and egg and form a new individual (thus further mixing the genetic instructions), an entirely unique organism is created unlike any other existing organism. The variations or differences caused by meiosis are very important to evolution, since the process of natural selection (the process of survival and reproduction of organisms that are best suited to their environment) needs genetic variety from which to "select." If there were no differences, there would be no evolution.

In humans, meiosis occurs in the gametes or sex cells (sperm and egg). In males, the process of gamete production is known as spermatogenesis. During this process, each dividing cell in the testes produces four functional sperm cells, all basically the same size. In contrast, the female process of producing eggs, called oogenesis, makes four eggs, only one of which survives. This is because nature gives all the necessary cytoplasm (living material) and organelles (structures with particular functions) to only one egg, thereby increasing its chances of survival should it become fertilized.

[*See also* **Cell Division; Fertilization; Reproduction, Sexual**]

Membrane

A cell membrane or plasma membrane is a thin barrier that separates a cell from its surroundings. It also keeps the cell's cytoplasm and organelles on the inside. Membranes are selectively permeable, meaning that some things can pass through the membrane and some cannot.

All cells have a cell membrane that is a thin but double layer of molecules that surrounds it. This membrane acts as a barrier and helps protect the cell while controlling the movement of substances in and out of

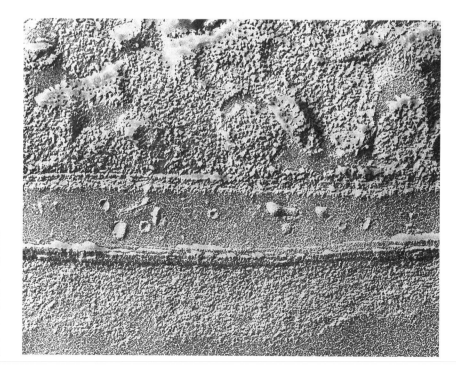

A freeze fracture image across the cell wall and membrane of a blue-green alga. (Reproduced by permission of Phototake NYC. Photograph by Dr. Dennis Kunkel.)

the cell. The membrane also allows a cell to maintain a constant internal environment, despite changes in its external environment. It is able to do this because of the semipermeable nature of its layers that regulate the passage of all substances going through it. The membrane is able to keep things out that it does not want or need while allowing in what it must have. A membrane surrounds not only the entire cell, but each organelle or specialized structure inside the cell also has a membrane around it. For both the cell and its organelles, the membrane is a place of constant activity. Although this membrane is extremely thin, it is very strong and can heal itself if broken. Examined very closely, a membrane is like a mesh bag whose little, square holes are small and strong enough to hold a dozen oranges or five pounds of potatoes, but which will also let water flow out or in completely. The bag is semipermeable, since it keeps certain-size things in (oranges) but let things of another size (water molecules) pass through. Most cell membranes are permeable to oxygen and water but not to large organic molecules like proteins.

One way that substances move through a membrane is by a process called passive transport. Passive transport involves no use of energy on the part of the cell, since certain substances are able to move freely in or out of it. Diffusion and osmosis are forms of passive transport since the cell does not need to use any energy to move substances. In diffusion, molecules of a substance spread themselves out more evenly from an area of high concentration. Substances like carbon dioxide, salts, and oxygen move in and out of cells this way. In osmosis, water molecules move from an area where they are crowded and cross a membrane to where they are less crowded. This process stops of its own accord when the solutions on either side of the membrane are at equal strength.

Active transport is a different way that a cell moves molecules in or out through its membrane, and it involves the use of energy. Active transport occurs when a cell wants to bring in more of a substance than will enter via passive transport. This happens when a plant's nearly-full root cells want to store even more minerals, and must move them from an area where they are less crowded to one where they are more crowded (inside the cell). In order to pack in more molecules, the cell uses carrier molecules that literally carry the desired molecule to a membrane slot into which it fits and then forces it into the cell. This process requires that the cell uses its own energy. Cell membranes are much more than walls or barriers that hold a cell together and keep it separate from its environment, since membranes control the movement of substances into and out of a cell.

[*See also* **Cell; Diffusion; Osmosis**]

Mendelian Laws of Inheritance

The Mendelian laws of inheritance laid down the basic principles of genetics. They state that characteristics are not inherited in a random way, but instead follow predictable, mathematical patterns. Mendelian laws were formulated by Austrian monk and botanist (a person specializing in the study of plants) Gregor Johann Mendel (1822–1887) in 1865, but went unnoticed for nearly a half century.

Before Mendel, many scientists had realized that certain traits, or characteristics, were passed on from one generation to the next, but in the middle of the nineteenth century, no one had any idea about where to begin to discover what controlled them or how or why these traits were passed on. In 1857 Mendel was able to combine his interest in both botany and mathematics by undertaking a long-term study breeding garden peas. For the next eight years, Mendel was able to conduct a thorough scientific study of how traits pass from one generation to the next.

By using ordinary garden peas—like those we eat today and call sweet peas—Mendel was able to easily breed for what are called "pure traits." This means that a self-pollinated (plants that contain both male and female reproductive organs and are able to transfer pollen between these parts) plant with pure traits will always produce offspring like itself. For example, a purebred plant that produces yellow pods will always produce yellow pods. Mendel then selected pea varieties that differed in single traits (such as height or pod color), and then he crossed them with plants that had a different trait (crossing tall plants with short, or yellow pods with green). After crossing a pure tall with a pure short, he would record the number of each type harvested and save the seeds produced by each plant for later planting, recording, and study.

While Mendel was conducting these careful experiments, neither he nor anyone else had any idea that such things as chromosomes (coiled structures in a cell's nucleus that carries the cell's heredity information) and genes (basic units of heredity) existed, although he would eventually decide that plants contained something he called "factors" and "particles of inheritance." He came to this conclusion because of the pattern of results he eventually saw. The very first thing that Mendel discovered after crossing a pure tall plant with a pure short one was that it did not result in the production of medium-size offspring. Instead, in the first generation, all the plants were tall. However, after allowing these plants to self-pollinate, he saw that the next generation produced plants that were a mix of tall and short. In fact, three-quarters were tall (which he called a dominant factor), and one-quarter were short (which he called a recessive factor). Mendel continued crossing hundreds of plants and kept careful records. Eventually he was able to state that a regular 3 to 1 ratio or pattern existed for the number of dominant versus recessive traits. This led him to realize that there must be laws or rules that make this mathematical ratio happen.

Continued work and study eventually allowed him to formulate what are now called the Mendelian laws of inheritance. He stated correctly that the characteristics of an organism are passed on from on generation to another by definite particles (which he called factors and we call genes). These genes exist in pairs, which are really different versions of the same genetic instructions. In this pair, one of the two factors comes from the male parent and the other comes from the female parent (each contributes equally). Finally, traits do not blend but remain distinct, and they combine and sort themselves out according to fixed rules. Mendel also stated

Two labeled diagrams showing Mendel's first law of inheritance, the law of segregation. (Illustration by Hans & Cassidy. Courtesy of Gale Research.)

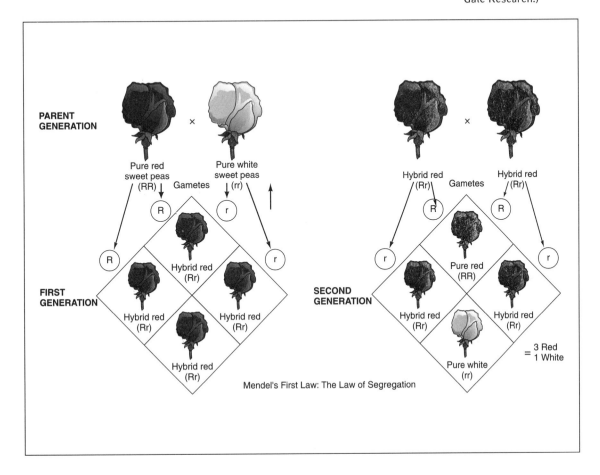

Mendel's First Law: The Law of Segregation

that dominant genes always have an effect on an individual, but that two recessive factors have to be present before they are expressed. Recessive factors can therefore be present in an individual but not have any effect on its characteristics.

Mendel published his findings in an obscure journal, and although his laws had laid the foundation for the new science of genetics, his work remained unknown for nearly two decades after his death. In 1900 Mendel's work was separately discovered by three different botanists (in three different countries) who realized that Mendel had discovered the laws of genetics long before they had. Although each published his own version of these laws, each man cited Mendel as the real discoverer. All three honorably stated that their work was merely a confirmation of what Mendel had accomplished in 1865.

[*See also* **Chromosomes; Gene; Genetics; Inherited Traits**]

Metabolism

Metabolism refers to all of the chemical processes that take place in an organism when it obtains and uses energy. Metabolism can be divided into two major phases in which substances are broken down and other substances are made. All organisms conduct both phases constantly.

If the body of a living thing is thought of as a machine, then its metabolism is similar to a running motor. In an organism, however, the motor is not only never turned off, but it is able to monitor itself and make adjustments according to internal and external changes that are always taking place. The job of the entire organism, just like the job of every living cell, is to conduct metabolic reactions continuously. These reactions all center around the processing and use of energy. Such reactions are needed by cells and the entire organism to constantly fuel itself, repair itself, and grow. The entire range of these chemical reactions make up an organism's metabolism.

TYPES OF METABOLISM

Metabolism can be divided into two phases or categories: catabolic metabolism (catabolism) and anabolic metabolism (anabolism). Catabolism is also known as destructive metabolism. It involves the breaking down of the molecules in nutrients taken in and the release of the energy they contain. Through catabolic processes, complex compounds like fats, carbohydrates, and proteins, are degraded (broken down) into simple molecules so their energy can be released. Catabolism takes place in the body when food is digested. However, if the system needs energy and no food is available, catabolism can also break down the body's stored fat and protein.

Anabolism is also called biosynthesis or constructive metabolism and can be considered the reverse of catabolism. By its different names, it is apparent that these types of chemical reactions involve the synthesis or the making of essential, complex molecules from simpler components. Anabolism is the body's building-up phase in which it uses the complex substances it has just formed for growth and overall body maintenance. For instance, by combining amino acids (the building blocks of protein molecules), the body's cells can form structural proteins and use them to repair and replace worn-out tissues. It can also form functional proteins such as enzymes to speed up chemical reactions, antibodies to fight disease, and hormones to regulate body processes.

The remarkable thing about an organism's metabolism is that its two phases of chemical reactions are going on constantly, stopping only when the organism dies. This constant and highly complex level of chemical activity has built-in control mechanisms that regularly monitor and adjust to all sorts of changing conditions. Hormones control metabolism, and thyroxine, which is secreted by the thyroid gland, determines the rate of metabolism (meaning the rate at which the body uses up energy to perform a certain metabolic function). The pancreas determines whether anabolism or catabolism is being performed, and then releases either insulin or glucagon. Since eating causes the level of glucose in the blood to rise, the pancreas responds by releasing insulin which, in turn, starts the process of anabolism. If the glucose level is low, the pancreas releases the hormone glucagon, which triggers the catabolic processes.

THE BASAL METABOLISM RATE (BMR)

A person's metabolic rate, or the rate at which it releases energy, is influenced by a number of factors, such as an individual's age, sex, level of activity, general health, and hormone levels. Because an individual's metabolic rate can provide a doctor with a great deal of information, it often needs to be measured. Since almost all of the energy used by the body is eventually converted to heat, the metabolic rate is usually calculated by measuring the amount of heat loss an individual displays during basal (resting) conditions. A person's Basal Metabolism Rate (BMR), therefore, measures the amount of energy the body consumes in performing its maintenance operations, like normal breathing, heart beating, and minor movements while it is at rest (but not asleep). A person's BMR is then judged to be normal or abnormal by comparing it to standardized

SANTORIO SANTORIO

Italian physiologist Santorio Santorio (1561–1636) founded the modern study of animal metabolism (all of the chemical processes that take place in an organism when it obtains and uses energy). He introduced the idea of quantification (number or quantity of) and measurement into the study of human physiology (the study of how different processes in living things work) and used mathematics and experimentation as his tools. He was an original thinker who was far ahead of his time.

As part of a tradition that was not unusual for Italy, Santorio was given the same first name as his last. Even when he became well-known and his name was Latinized as Sanctorius Sanctorius, it was still the same double name. He was born in Capodistria (what is now Croatia), and his father was an official in the Republic of Venice. Santorio was soon sent to Venice where he was educated by tutors. He began the study of philosophy and medicine at the age of fourteen, and obtained his medical degree from the University of Padua in 1582. For the next decade or so, it is unclear whether he spent time in Poland working as a physician for the king. Biographers disagree, but it is known that he was often in Venice during this period. In 1611, however, he was appointed professor of theoretical medicine at the University of Padua on the strength of a medical book he had written in 1602. At Padua he was a very popular lecturer, and students came from all over Europe to attend.

It was in this 1602 medical book that he first began writing about how important it was to measure things exactly when doing physiology. He also stressed that close and careful observation was equally important. Ten years later he published another medical book based on his more extensive

rates or rates that reflect the average BMR of healthy individuals of various ages.

Typically, the BMR of males is higher than that of females, and both sexes have a lower BMR as they age. A recent discovery relates metabolism to the aging process, since animal studies suggest that there is some connection between substantially reducing calorie intake and living longer. It has yet to be determined what role metabolism plays in this phenomenon.

Finally, since enzymes and hormones play such a key role in every metabolic process, a glandular problem or a genetic fault that affects the production of an enzyme can result in major metabolism problems. Such conditions as diabetes and Addison's disease throw a person's metabo-

research, and in it he described his use of a type of thermometer, probably invented by Italian physicist and astronomer Galileo Galilei (1564–1642), which he adapted to measure the warmth of the body. He was thus the first physician to use a thermometer on a person and to write about it. This could be said to be the first clinical thermometer. In yet another book written in 1614, Santorio described his experiment weighing a human being every day to determine the influence of everything that went into and came out of a body, including perspiration. He even designed an apparatus that had a chair built onto a large scale and was able to prove with this that people lost weight by the evaporation of their perspiration. His ability to measure all things related to the body was further improved when he invented a device to measure a person's pulse rate. For more than twenty-five years, Santorio performed experiments on more than 10,000 subjects, using scales and similar measuring instruments. This led modern biologists to call him "the father of the science of metabolism." If metabolism can be described as all of the chemical processes that take place in a living thing, then metabolism is indeed what Santorio was pioneering.

However, his work did not have any great impact on the science of his era, and many scientists believe it is because he was simply too far ahead of his time. Nonetheless, as science progressed, his modern ideas came to be more understood and appreciated. Santorio always stressed the importance of measurement, facts, and solid information, and regularly argued against such unscientific practices as astrology (the use of stars and planets to predict their influence on human affairs). He was in favor of applying all the new tools and instruments that science had in its possession to the study of the human body. Although he did not know the word for what he was doing, he was studying what is now called human metabolism.

lism off severely, while other people simply cannot tolerate or process certain foods, like milk.

Metamorphosis

Metamorphosis comes from a Greek word meaning "transformation," and is the term biologists use to describe the extreme changes that some organisms go through when they pass from an egg to a adult. A caterpillar turning into a butterfly is an example of complete metamorphosis. Metamorphosis often gives an organism some type of competitive advantage and usually occurs in organisms with short life spans like amphibians, some fish, and various invertebrates (animals without a backbone), especially insects.

Metamorphosis means much more than a physical change, however. In the process called complete metamorphosis, major differences occur not only in outward appearance but in an organism's internal organs and processes as well. Sometimes, an adult or mature organism has an entirely different set of cells and organs compared to what it had in an earlier stage of its development. Many organisms look different as adults compared to what they looked like when just born or very young. For example, a crowing adult rooster does not resemble the chirping, yellow ball of fluff that it was as a chick. This is not an example of metamorphosis, but simply a difference in size and an elaboration of certain characteristics. For metamorphosis to occur, an organism must go through at least three stages of development during which it changes radically both inside and out.

Metamorphosis does not happen in the life cycles of what are called "higher animals," such as dogs, cats, or human beings. Rather, metamorphosis occurs only in certain "lower animals" like ants, butterflies, sea urchins, and frogs. During the most dramatic example of complete metamorphosis, an organism passes through four distinct stages of development and ends up being a completely different type of organism.

The butterfly is probably the best-known example of an organism whose life cycle undergoes complete metamorphosis. In its embryonic or egg stage, it is deposited as an egg on a green plant that will serve as its food when it hatches. Its second stage is its feeding time and is called its larva stage. A larva is an insect that is in its wormlike stage. This stage begins when the egg hatches. In the case of a butterfly, what emerges from the egg is a caterpillar. It may be hairy or smooth and have distinct markings or little color according to what species of butterfly its parent was. At this point in its life, the larva or caterpillar has legs, and chewing mouthparts but no wings, and is best described as an eating machine. It looks more like a worm than an insect that will fly. The caterpillar eats and continues to grow and go through several molts (the shedding of its outer skin). As it grows, its body parts continue to be rearranged and modified and it finally matures into an adult caterpillar. At this point, it is ready to enter the third stage of complete metamorphosis called its pupa stage or cocoon stage. When ready, the larva attaches itself to a branch or a twig and forms a protective covering around itself. Butterflies form a hard, shiny shell called a chrysalis that hangs suspended from a twig, while moths spin their coverings out of silk and wrap themselves almost flat against the twig. It is during this pupal or resting stage that the larva

Opposite: Beginning with the larva in the lower left-hand corner, this illustration shows the transformations a gypsy moth goes through during metamorphosis. (Illustration courtesy of The Library of Congress.)

changes into the adult butterfly it will eventually become. Inside the chrysalis or cocoon, most of the larva's cells are broken down and new tissues and organs begin to develop. Eventually, an entirely new and different organism is created using the raw materials left by the old one. When this transformation is finished, the chrysalis or cocoon breaks open and the adult butterfly emerges. After its shrunken wings stretch and fill with blood, they soon are strong enough for the butterfly to fly away. This is its final stage—called its adult stage. As an adult, the insect will eventually reproduce and lay its eggs, beginning the cycle and metamorphosis once more.

The often startling aspect about metamorphosis is how suddenly and dramatically one form of life is changed into another that seems completely different. A butterfly in its larva stage crawls about and eats leaves, while as an adult it flies from flower to flower and sips nectar. The same transformation occurs for some amphibians. An example is frog eggs that develop into swimming tadpoles and breathe with gills like a fish. After they change into an adult, nearly every organ has changed. The tadpoles' tails are absorbed into the legs of a frog and they breathe with lungs. For all organisms that go through metamorphosis (beetles, flies, ants, bees, and wasps as well as butterflies and frogs), their hormones begin the process and keep it going.

Unlike butterflies and frogs, some insects go through only a three-stage process (egg, nymph or larva, and adult) called incomplete metamorphosis. In this abbreviated version, the pupa stage is omitted and the changes are more gradual. Grasshoppers, crickets, cockroaches, and termites go through incomplete

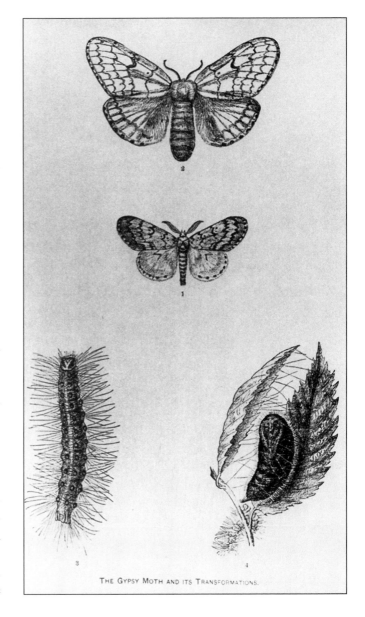

THE GYPSY MOTH AND ITS TRANSFORMATIONS.

metamorphosis. Even by undergoing only incomplete metamorphosis, it is believed that species that live entirely different lives as a young organism and as an adult have an advantage over those who do not. For example, a species that is slow-moving and limited in range in one stage of its life (such as a caterpillar) can suddenly move around quickly as an adult and can therefore lay its eggs far away in a more advantageous habitat. Another advantage is that overall competition is reduced, since at one stage an organism is seeking one certain type of food, and in a later stage it eats something entirely different.

[*See also* **Amphibians; Insects; Larva; Life Cycle**]

Microorganism

A microorganism is any form of life that is too small to be seen without a microscope. Also called microbes, these tiny organisms include bacteria, protozoa, single-celled algae, and fungi as well as viruses. Microorganisms are nearly everywhere and are essential to the production of certain medicines, foods, and drinks. They play a key role in nature's oxygen, carbon, and nitrogen cycles, but can also be harmful to humans.

The study of microorganisms is called microbiology. It was founded in the seventeenth century after the microscope was invented. Not until the Dutch naturalist Anton van Leeuwenhoek (1632–1723) saw what he called "little animalcules" (or little animals) with his own microscope in 1673, did science know of the existence of a subvisible world that was teeming with life. Despite this major discovery, no one knew where these microscopic forms of life came from. Most believed that the life forms simply sprang out of rotten wheat or from the soil. This incorrect notion of spontaneous generation was held until 1861 when the French chemist Louis Pasteur (1822–1895) was able to prove that spontaneous generation does not occur and that the air itself is full of microorganisms. Pasteur also discovered that fermentation was caused by microorganisms. (Fermentation is a process in which cells break down sugar or starch into carbon dioxide, ether, alcohol or lactic acid.)

Pasteur went on to make many other contributions and is considered the founder of the science of microbiology. He discovered that microorganisms are present in nonliving matter as well as in the air, and that some of these tiny organisms caused disease. He also showed how microorganisms could be killed by heat and how they could be manipulated for use in vaccines. In 1876, the German physician Robert Koch (1843–1910) was able to demonstrate that a particular bacterium could cause a particular disease. He isolated the bacteria that produced the cattle disease anthrax, and he discovered the bacteria that caused tuberculosis and cholera. With the discovery of viruses just before the turn of the century, it was realized that although viruses are not living organisms (since they cannot grow and reproduce on their own), they are even smaller than bacteria and are physically microorganisms.

There are several types of microorganisms. Protists are a group of single-celled plantlike or animal-like organisms that have complex or eukaryotic cells. This means that their genetic material is contained in a nucleus that is bound by a membrane. Protists (kingdom Protista) are far more diverse than plants or animals and obtain their food by simply absorbing it from their environment since most live in water. They also have different means of getting about, such as flagella (tails) and cilia (hairs). The well-known euglena is a protist, as are all single-celled algae and the many differently shaped diatoms that form glasslike shells. Bacteria are another group of microorganisms (too small to be seen with the naked eye), that belong to the kingdom Monera instead of Protista. Among the most abundant life forms on Earth, bacteria are single-celled organisms that have a cell wall but no nucleus (meaning that they are prokaryotic and not eukaryotic). Some bacteria feed on dead matter and play an important role in recycling nutrients, while others cause disease. Fungi are microorganisms that make up their own kingdom (Fungi). The smallest fungi can be microscopic and single-celled. Like protists, fungi also absorb their food from their environment. Finally, viruses are microorganisms by virtue of their size. They are not considered living organisms because they are parasites and can reproduce only by taking over their host's cellular machinery. They are the tiniest of all the microorganisms and can only be seen with an electron microscope.

Microorganisms play a major role in the environment. As recyclers, they break down many substances into usable forms for plants and animals. Without microorganisms the world would be full of waste. Microorganisms are also crucial to several key industries. For example, the production of antibiotics, vaccines, beer, cheese, wine, and bread would be impossible without microorganisms. Although sometimes beneficial, microorganisms also can cause diseases. Some of these diseases are less dangerous than in the past because scientists have been able to develop cures. However, other diseases continue to be harmful as microorganisms adapt and mutate in response to treatment.

Knowledge of microorganisms has allowed biologists to use them as experimental models to study the chemical processes of more complex organisms. Microorganisms have been crucial to our knowledge of deoxyribonucleic acid (DNA) and have proven to be an essential tool of genetic engineering in which biologists experiment with the genetic code of living organisms. Altogether, microorganisms are not only essential to life on Earth, but also help scientists solve problems in medicine, agriculture, industry, and the environment.

Microscope

A microscope is a scientific instrument that magnifies objects that are too small to be seen by the naked eye. As one of the most important scientific tools ever invented, it is especially significant to the life sciences, since it made possible the discovery of an entirely unseen world of microorganisms. Today's increasingly powerful and highly specialized microscopes can achieve magnifications of a million times or more.

The earliest types of magnifiers were probably globes of water-filled glass or chips of transparent rock crystal used by the Romans. The first microscope could not be invented, however, until the first lenses were devised for use in eyeglasses sometime around the year 1300. These first eyeglasses or spectacles were made with convex lenses (curved inward) that helped farsightedness (the inability to see objects up close). By 1500, it is known that concave lenses (curving outward) were crafted to help with myopia, or nearsightedness (the inability to see object far away). Lenscrafters had learned that by grinding any clear glass or crystal into a certain shape, usually with the edges thinner than the center, a magnifying effect was achieved. The first real microscope was therefore a single, handheld lens, and it is called a simple microscope. Today, we would call it a magnifying glass. The individual most identified with the improvement and use of the simple microscope is the Dutch naturalist Anton van Leeuwenhoek (1632–1723), whose secret grinding, polishing, and mounting techniques allowed him to achieve possibly as much as 270 times magnification. Beginning in the 1670s, he examined mainly biological specimens and was the first to observe spermatozoa (male sex cells), red blood cells, and bacteria.

The typical microscope used today is a tubelike instrument with a lens at its top and bottom. It is called a compound microscope because it has more than one lens. This device is believed to have come about in Holland near the end of the sixteenth century when the telescope was invented there at the same time. Apparently, it was soon realized that a telescope could be used as a microscope when reversed. The first compound microscopes were, therefore, two lenses housed in a long tube (in which the enlarged image produced by the first lens is further magnified by the second one). The first scientist to improve the compound microscope and to put it to real scientific use was the English physicist, Robert Hooke (1635–1703). In 1665, Hooke published *Micrographia,* which contained excellent drawings of what he had observed with his improved microscope. Hooke was the first to use a microscope to observe the structure of plants (actually thin slices of cork), finding that they consisted of tiny walled chambers that he called "cells." After Hooke, there were minor improvements in microscopy until the mid-1800s, when the German physicist Ernst Abbe (1840–1905) collaborated with the German optician Carl Zeiss (1816–1888), and produced high-quality lenses with no blurring or distortions. Later developments resulted in the basic microscope with built-in illumination (lighting) that is used today in schools and small laboratories. These generally have a magnification of up to 400 power (times).

A labeled diagram showing the components of a modern compound light microscope. The invention of the microscope was truly revolutionary since it allowed scientists to see objects unable to be seen with the naked eye. (Reproduced by permission of Carolina Biological Supply Company/Phototake NYC.)

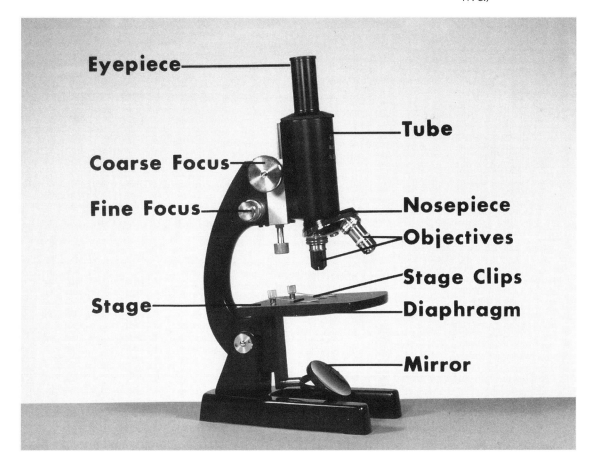

ROBERT HOOKE

English physicist (a person specializing in the study of energy and matter and their interactions) Robert Hooke (1635–1703) was one of the earliest and greatest of the microscope pioneers. His microscopic studies of insects, feathers, and fish scales are both beautiful and accurate, and he published the first book dedicated to microscopy. He is responsible for first using the word "cell," which later would become the cornerstone of microbiology (the study of microorganisms).

Born on the Isle of Wight in England, Robert Hooke was born sickly and with a backbone that did not grow straight. As a child prodigy (exceptionally smart) Hooke impressed everyone with his mechanical gifts. He built elaborately complicated toys as a child, and as a young man he attended Westminster School and later Oxford University. At Oxford, his abilities caught the eye of the great English physicist and chemist, Robert Boyle (1627–1691), and Hooke quickly was made Boyle's assistant. It is known that it was Hooke who designed and built an improved air pump that Boyle used to establish his gas laws. When Hooke was made a member of the Royal Society in 1663, he also became its "curator of experiments," which allowed him access to the society's facilities. He remained in this position for the rest of his life, and was able to pursue whatever interested him scientifically.

As a man of wide talents, Hooke's scientific interests were even broader, and he went on to make contributions in physics, astronomy, architecture, microscopy, and biology, among other fields. Although Hooke was not the first to experiment using a microscope, he was the first to dedicate an en-

Today's school and lab microscopes are called compound light microscopes because they let light pass through the object being studied and then through two or more lenses. The lenses enlarge the image and bend the light toward the eye. Such a microscope has two lenses: an objective lens and an ocular lens. The ocular is also called the eyepiece and is what you look through. The objective lens, sometimes only called the objective, magnifies the object just below it on a slide. If the objective lens has a power of 50X (magnifying an object 50 times), and the ocular has a power of 10X, then together they have a total magnification of 500X (10X times 50X). Such a magnifying power would allow a cell to be easily observed.

While such a microscope is adequate for schools and modest labs, greater magnification is often needed for more advanced research. In 1931 the electron microscope was invented. Using the knowledge that beams of electrons (particles that make up a single atom) could be focused us-

tire book to microscopy. In 1665, he published his *Micrographia,* which was written in English despite its Latin title. This work contains descriptions and illustrations of the structures of insects, fossils, and plants in never-before-seen detail. His drawings of tiny insects, parts of bird feathers, and even fish scales are both artistically beautiful and scientifically accurate. The biological discovery for which he is best remembered is the porous (having pores or holes) structure of cork. When he took a thin slice of cork and put it under the compound microscope that he had built himself, he noticed that it was made up of tiny rectangular holes that he called "cells." It was an appropriate name for these little boxes, or empty rectangular structures, since the word usually meant a small room (like a jail cell). In fact, what he was seeing were the now-dead remnants of once-living structures that had been filled with fluid. That is, he actually was viewing what had been cells. Hooke's word "cell" came to be adopted by biologists once they were able to observe living structures under a microscope. Ever since, the word and the concept it stands for has become one of the cornerstones of biology. Hooke's *Micrographia* is recognized today as containing some of the best microscopic views of nature. In addition to his microscopic studies of insects and plants, he studied fossils a great deal, which led him to offer some early ideas about what is now realized to be evolution (the process by which living things change over generations). Hooke was described by some as quarrelsome, miserly, and a hypochondriac (someone who believes that are always sick), and he engaged some of the best minds of his time in some terrible feuds. Yet his contributions to biology are, in the full range of his life, but a small part of what he accomplished and contributed to other fields of science.

ing a magnetic field the same way a glass lens focuses light, scientists built an electron microscope a thousand times more powerful than a regular compound microscope. Today there are two types of electron microscopes: the transmission electron microscope (TEM) and the scanning electron microscope (SEM). In a TEM, electrons pass through the object and cannot be used to magnify living things. In a SEM, electrons are bounced off the surface of the object, meaning that it is possible to examine things that are alive. Although the SEM cannot magnify things as greatly as the TEM, it produces a more three-dimensional image. Both microscopes display their magnified images on a video screen.

Besides the TEM and SEM, other more advanced type of microscopes include the phase contrast and dark-field microscopes. The phase microscope contrasts light waves that pass through a specimen with those that do not, making for a sharper contrast. The dark field microscope makes an object appear bright against a dark background. Finally, a scanning optical microscope uses a laser to obtain pinpoint illumination, while the atomic-force microscope uses a diamond-tipped probe that moves across the surface of the specimen and is able to "see" individual living cells without damaging them. A microscope extends the sense of sight to an incredible degree, and while it has been an important instrument to all of science, it is an essential tool for the life sciences.

Migration

Migration is the seasonal movement of an animal to a place that offers more favorable living conditions. Migrations can be long or short distances and are usually annual, involving a round trip. Although there are many risks involved in migrating, they are usually outweighed by the benefits.

Although some people consider the one-way, permanent relocation of an animal to be a migration (such as when a species is driven from its natural habitat by a sudden change in living conditions), most scientists consider migration be a periodic, regular occurrence in which certain animals make a regular, round-trip journey to the same place. A good example is the Arctic tern, the animal that makes the longest known migration. This remarkable bird leaves its arctic home as winter approaches and flies south to Antarctica to take advantage of its "summer." When winter approaches in the south, the tern returns north to take advantage of the arctic "summer" there. Altogether, it flies about 25,000 miles in a single year.

However, migrations do not necessarily have to be as long as the Arctic tern's to be considered real migrations. Some species, like the grizzly bears that frequent the Rocky Mountains, make "vertical" migrations in which they migrate to high-altitude tundra meadows in early summer where they feed. Later in the season they come down to feed on plants growing in lower-altitude valleys and meadows. However, most animals usually make longer migrations. The North American bison once migrated in enormous herds from their summer range in the northern prairie to their winter habitats further south. Certain fish can also be long-distance migrants, such as the Atlantic salmon. These salmon hatch in large rivers and migrate as juveniles to the open ocean. There they feed for several years until they are sexually mature, after which they migrate back upriver to the stream where they were born in order to breed.

Most migrations take place on a yearly basis. The effort that an animal has to make in order to do this is usually considerable, making it an "expensive" activity. This means that it costs an animal great amounts of energy to move itself from one place to another. Migrating also can involve substantial risk, since it becomes a time when the animal is exposed to stress and predators as well as to the possibility of getting lost. Despite this, migration is an adaptive behavior that is carried out because it confers some sort of benefit on the animal that does it. These "benefits" usually outweigh the "costs."

REASONS FOR MIGRATION

Among the benefits of migration are a greater abundance of food and more favorable weather conditions. Some animals actually follow their food. For example, the dramatic migration of millions of monarch butterflies from eastern Canada and the United States to Mexico is believed to be related to the availability of milkweed plants on which these butterflies feed and lay their eggs.

Certain animals migrate to specific areas to mate, lay eggs, or give birth, while others migrate in order to raise their young under favorable conditions. For example, many birds migrate where there are favorable nesting places and plentiful insects to feed their ravenous young. Humpback whales migrate from their polar waters and give birth in warm tropical waters. In these waters the young can easily thrive and can build up layers of insulating blubber to fight the extreme cold of their future feeding grounds.

Whatever the reason, it appears that migration in certain animals is instinctive or inherited behavior. This means that the migrating animal reacts to certain "cues" and acts out a pattern of behavior that all the generations before it have done. Studies have shown that different animals respond to different cues. For some, it is the declining hours of daylight. Others respond to temperature or weather changes, and some react specifically to changes in rainfall amounts. It is believed that these changes cause an animal to respond because they trigger the release of certain hormones that are the real stimulants to migration.

There is mystery in exactly how these different species manage to find their way, sometimes over great distances. It is known that some, like salmon, are guided by their sense of smell, while others, like whales, use echolocation or bounce sounds off objects to orient themselves. Some, like the caribou, simply follow well-trodden trails or use landmarks. Many birds use the positions of the Sun and stars to navigate or are believed to be able to sense Earth's magnetic fields. Overall, migration is an animal's response to seasonal changes in the availability of the things it values, such as food, shelter, and mates. If their existing habitat cannot provide those things, then these animals will physically relocate in order to find them.

Mitochondria

Mitochondria are specialized structures inside a cell that break down food and release energy. If a cell is like a tiny chemical processing plant, then the mitochondria are the power plants of the cell. Without mitochondria, a cell with a nucleus cannot use oxygen and cannot live. Mitochondria are found only in eukaryotic cells (those with a nucleus) and are not found in prokaryotic cells (those without a nucleus).

Mitochondria (singular, mitochondrian) are described as organelles. An organelle is a tiny structure inside a cell that is controlled by the nucleus and has a specific function to play in maintaining the life of the cell. For a cell, mitochondria play a most important role in carrying out aerobic respiration. In other words, mitochondria break down the food a cell takes in and release the energy it contains. Examined under a microscope, mitochondria appear as oval- or sausage-shaped structures that have a double membrane. The outer membrane is smooth and permeable to certain enzyme molecules, meaning that molecules of a certain size

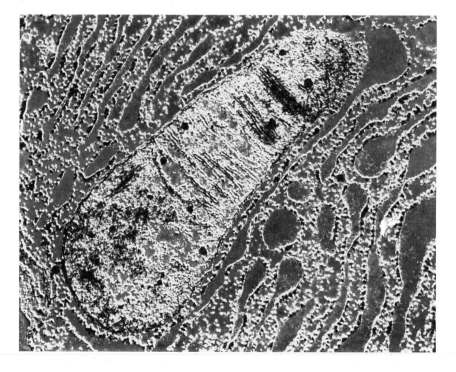

A colored transmission electron micrograph of a mitochondrian, and rough endoplasmic reticulum. The mitochondria are considered the "powerhouse" of the cell since they break down food and release energy. (©1996 by SPL/Secchi-Lecaque/Roussel-UCLAF. Reproduced by permission of Custom Medical Stock Photo, Inc.)

can flow through its walls. The inner membrane consists of many folds that allow it to have an increased surface area and make it able to pack in many more enzymes than would be possible without the folds. A eukaryotic cell typically has anywhere from a dozen to a thousand mitochondria, and animal cells usually have more mitochondria than plant cells. Mitochondria are especially abundant in cells whose functions have high energy demands.

As the "powerhouse" of the cell, mitochondria carry out respiration. Respiration is the chemical process that breaks down food to release energy. Two key ingredients are necessary for respiration—sugar and oxygen. The sugar that cells use is usually in the form of glucose, which contains a great deal of stored energy. Oxygen is used to get the energy out of the glucose. This process takes place in the folds of the mitochondria where a substance known as adenosine triphosphate or ATP is created. ATP is the molecular storage form of energy that mitochondria produce. When a cell needs energy, it draws on the stored ATP, which releases the energy.

Here is a simplified example of how the human body uses mitochondria to release energy from food. For an organism like the human body, most of its energy comes from mitochondria, which are the body's power plant that burns the fuel to produce the energy to run it. The food a person eats is the fuel that is "burned" in the body's furnaces (mitochondria). The ATP produced by this process is the energy or "electricity" produced by the power plant (which powers the cells in the body). When a person breathes, he or she is taking in oxygen for the mitochondria to use to release the energy the body needs. This process is similar in all organisms, since all organisms must break down food to get energy. As a result, all life forms have cells that contain mitochondria.

[*See also* **Cells; Organelle**]

Mollusk

A mollusk is a soft-bodied invertebrate (an animal without a backbone) that is often protected by a hard shell. Most mollusks live in water and make up the second largest group of invertebrates (next to arthropods). Mollusks are represented by such diverse invertebrates as clams, slugs, snails, and octopuses.

There are about 100,000 species of mollusks, most of which live in the water. All are soft-bodied, nonsegmented, and usually enclosed in some sort of covering made of calcium carbonate. Whether a land snail or an underwater clam, all mollusks share certain traits. They all have a "mantle," or a covering of tissue that protects their internal organs. Many species secrete a substance that forms a hard shell, protecting them from predators. Their digestive systems are made up of a mouth, throat, gullet, stomach, intestine, and anus. Mollusks that live underwater use gills to get their oxygen, while land-dwelling mollusks have lungs. All have a two-chambered heart that pumps blood through vessels that branch throughout their bodies. The mollusk nervous system consists of two pairs of nerve cords. Some mollusks have eyes and other sense organs. Most mollusks also have a muscular "foot" that is used for slow crawling or digging. In an octopus, this foot is divided into arms or tentacles lined with suction cups.

There are at least eight species of mollusks known, but most fit into three main groups—gastropods, bivalves, and cephalopods—that are grouped according to the shape of their muscular foot. The name gastropod means "stomach-foot" and describes those mollusks, like snails, that

This tree snail, hiding in its shell, is part of a group of mollusks called gastropods because it has a sucker-like foot. (Reproduced by permission of Field Mark Publications. Photograph by Robert J. Huffman.)

have a sucker-like foot. This is the largest group of mollusks and is best represented by the common garden snail. Most gastropods, excluding slugs, have a single shell often shaped like a spiral, or coil. They also have a large, flat foot that they use to slide along on top of the mucus they secrete. This is why they leave a visible, slimy trail behind them. A gastropod has a distinct head with tentacles that can move up and down like a periscope and act as eyes or sense organs. A snail also has rows of tiny teeth. Not just confined to land, there are many ocean-dwelling gastropods like periwinkles, abalones, and sea slugs.

A bivalve is a mollusk that has a two-part shell joined by a hinge. Bivalves are water animals and are best represented by clams, oysters, and scallops. All bivalves are filter feeders. They filter water through large gills that catch bits of food and absorb oxygen. To do this, however, they must keep their shells open, and as a result, expose their vulnerable soft body. A mollusk, like a clam, moves on the ocean's bottom by thrusting its large, hatchet-shaped, muscular foot between the open shells, thus pushing it along slowly. Unlike gastropods, bivalves have no head, and their sense organs are not well-developed. Bivalves do have a powerful adductor muscle that they use to clamp their shells tightly together.

Members of the cephalopod group of mollusks, like the octopus and squid, are free-swimming and have no shell. Unlike typical mollusks, they are fast-moving hunters and prey on other animals. An octopus has a sharp beak, like a parrot's, that it uses to rip its food and break open shells. In squid and octopuses, the mollusk foot has evolved into long arms, or tentacles, around the head. They use these tentacles to capture fish. In some ways, cephalopods resemble vertebrates because they have a solid internal support structure, like a skeleton, and a large head with eyes, as well as a good-sized brain. Cephalopods move by a form of jet propulsion in which they forcefully squirt water. They also can eject a cloud of dark-colored chemicals that covers their escape as they zoom away. A giant squid may extend its tentacles as far as 60 feet (18.29 meters) and has the largest eyes in the animal kingdom—up to 1 foot (0.3 meters) wide. Some squid and octopuses can change color like a chameleon. The only cephalopod with a shell on the outside of its body is the nautilus whose shell is divided into chambers. It lives in the largest chamber and adds another when it grows. It fills the empty chambers with a gas that makes it easier to swim. All three groups of mollusks reproduce sexually through the union of male sperm and female eggs. Unlike gastropods and cephalopods, however, bivalves are not necessarily either male or female. Instead they may contain both male and female organs.

[See also **Invertebrate**]

Monerans

Monerans are a group of one-celled organisms that do not have a nucleus. Along with Protists, Fungi, Plants, and Animals, Monerans make up the five kingdoms of living things. As one of the first life forms to evolve, they are today the most abundant living organisms on Earth. Monerans are found throughout the world and can live in freezing as well as extremely hot conditions.

Monerans belong to the kingdom Monera, which, unlike other kingdoms, is made up of only one member—bacteria. Monerans are usually microscopic life forms, and although some are smaller than viruses, others can be seen by the naked eye. They live not only on Earth, from hot springs to frozen wastelands, but inside other organisms as well. Nearly all multicelled plants and animals act as hosts to Monerans. Monerans are so abundant that it is estimated that the number of bacteria found in the human mouth would outnumber all of the people who have ever lived. They have left no fossil record that scientists can learn from, yet are believed to be among the oldest type of organisms still thriving. Unlike other living cells, Monerans are prokaryotic, meaning that they have no nucleus or any organelles (tiny structures inside a cell that have certain functions) inside their walls. Instead, the material that is usually found in the nucleus is scattered throughout the cell. Monerans can also reproduce asexually by binary fission, meaning that a single cell can divide itself into two identical "daughter" cells. Monerans can reach maturity in a phenomenally short time (about fifteen minutes), so that they are able to rapidly mutate or adapt to a changing environment.

TYPES OF BACTERIA

For some time, Monerans were not considered to be a separate kingdom, but advances in molecular biology now suggest that this kingdom is made up of three different types of bacteria: the archaebacteria, the eubacteria, and the cyanobacteria. Some scientists argue that cyanobacteria (formerly called blue-green algae) are part of the eubacteria, and that archaebacteria should form their own kingdom.

Archaebacteria. Archaebacteria are unique, since some species live in such harsh places as boiling mud, hot springs, and extremely salty water, while others live in the intestinal tracts of some mammals. The archaebacteria called thermoacidophiles thrive in the recently discovered deep-sea volcanic vents where it is extremely hot.

Eubacteria. Eubacteria are considered to be "true" bacteria and are composed of types with which scientists are most familiar. Some make their own food as plants do, while others get their energy by fermentation (a process by which cells break down sugar and starch into energy). Others are considered dangerous parasites. The bacteria in the soil that are able to "fix" or capture nitrogen from the air are eubacteria, as are those that live in ticks and cause Lyme disease and Rocky Mountain spotted fever. These bacteria are also responsible for such transmissible diseases as syphilis and gonorrhea, as well as botulism and diarrhea. Although sometimes harmful, eubacteria also can be beneficial. For example, the bacteria *Escherichia coli* that normally lives in the gut of mammals produces enzymes that help with the digestion of fats. Eubacteria also are necessary for the production of cheese, yogurt, and other fermented milk products.

Cyanobacteria. Cyanobacteria are a group that make their own food by photosynthesis. Also called blue-green algae, they use the process that plants employ of capturing the energy of the Sun and changing it into simple food substances. Most live in or on the water's surface, either by themselves or in large clusters called colonies.

IDENTIFYING BACTERIA

Although all bacteria are prokaryotic, it is the cell wall that gives different types of bacteria their different shapes and allows biologists to identify them. The most commonly occurring shapes are round (cocci), rod-shaped (bacillus), and coiled (spirillum). The bacteria with a round shape called cocci (singular, coccus) often form chains and are usually found in the human body. An example is *Streptococcus mutans,* which inhabits the mouth and causes tooth decay. A single rod-shaped bacterium is called a bacillus (plural, bacilli). The species *Bacillus anthrax* causes the cattle disease anthrax. Finally, the corkscrew-shaped bacteria are like twisted spirals and are called *Spirillum* (plural, *Spirilla*) or *Spirochaete.*

Despite our familiarity with the many disease-causing bacteria, most bacteria are not only harmless to human beings, but some have become indispensable. Besides the bacteria that allow us produce cheese and wine, and naturally break down and recycle our waste and sewage, bacteria have been used in medicine to produce such modern miracle drugs as antibiotics (like erythromycin and streptomycin). More and more, bacteria are used in the biotechnology industry, which is genetically altering certain bacteria so they will produce important compounds like insulin that are difficult to make artificially. As a result, the Monerans or bacteria are an important and essential part of the life cycle on Earth.

[*See also* **Kingdom**]

Muscular System

The muscular system is composed of all those body tissues capable of contracting and relaxing and, therefore, of producing movements in its body parts. Muscles are able to produce movement by converting chemical energy into mechanical energy. This conversion occurs due to the ability of muscles to contract and relax. Animals use muscles not only to move about but to operate their many necessary internal processes such as blood circulation.

Since animals cannot make their own food as plants do, they must be able to move about to locate things to eat. Movement is therefore a trait shared by all animals, from a one-celled amoeba to a killer whale. Animals are able to move because they have a support framework (a jointed skeleton) that is moved by muscles, which push bones together and pull them apart. Muscles are made up of long fibers of contractile tissue, meaning they have the ability to contract or grow shorter by means of a chemical change, as well as relax or return to their normal length. Muscles are rich in blood vessels that bring them food and oxygen and take away their wastes. The harder the muscles work, the more blood is transported to them.

TYPES OF MUSCLES

There are three types of muscle found in the muscular system: skeletal muscle, smooth muscle, and cardiac muscle. Although all three contract in basically the same manner, they look very different when examined under a microscope.

Skeletal Muscle. Skeletal muscle is attached to bones and can be said to make up the flesh of an animal. It is also called striated muscle since it is made up of long, striped muscle fibers. Skeletal muscle is a voluntary muscle because it can be consciously controlled. A person can, therefore decide when and how to move his or her arms or legs to walk or run, or to move facial muscles to smile or frown. Most skeletal muscles are attached to bones and are connected at both ends by a tough, connecting tissue called a tendon. Tendons can be felt in the forearms near the wrist, in back of the leg near the knee, and in the back of the ankle. Skeletal muscles move bone by acting in pairs. One muscle, called the flexor, contracts and pulls the bones connected by a joint together. Another muscle, called the extensor, pulls, or moves the bones apart. Muscles, such as the flexor and extensor that work together but in opposing ways, are called an antagonistic pair.

Smooth Muscles. Smooth muscle is a second type of muscle, which actually appears smooth under a microscope. It makes up the walls of many

Opposite: Some of the estimated 800 muscles found in the human body. No exact figure is available because scientists disagree about which ones are separate muscles and which ones are part of a large muscles. (Illustration by Kopp Illustration, Inc.)

Muscular System

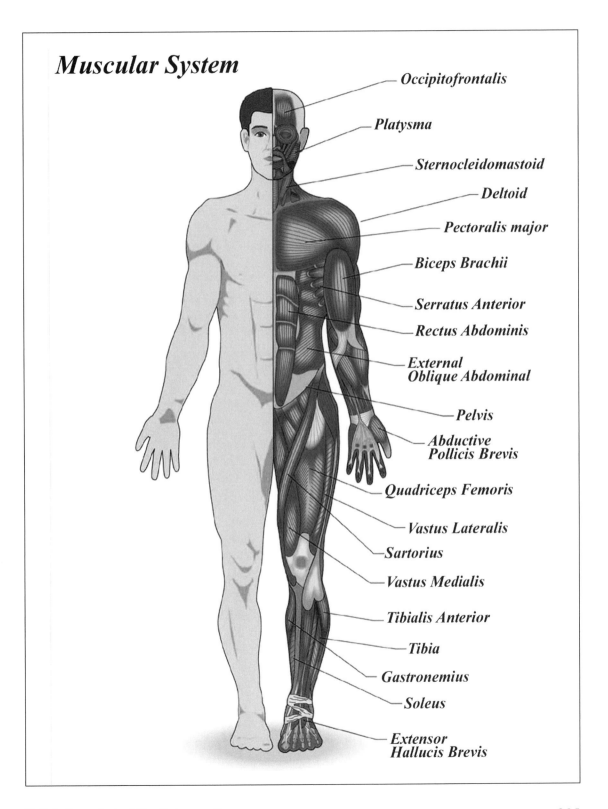

Occipitofrontalis

Platysma

Sternocleidomastoid

Deltoid

Pectoralis major

Biceps Brachii

Serratus Anterior

Rectus Abdominis

External
Oblique Abdominal

Pelvis

Abductive
Pollicis Brevis

Quadriceps Femoris

Vastus Lateralis

Sartorius

Vastus Medialis

Tibialis Anterior

Tibia

Gastronemius

Soleus

Extensor
Hallucis Brevis

organs inside the body, including the digestive tract, reproductive organs, bladder, arteries, and veins. Since it does not respond to a person's will or command—that is, it cannot deliberately be controlled—it is called involuntary muscle. Blood is pushed through the veins, and the lungs expand whether a person wants them to or not. Smooth muscles contract like any muscle, and although they do so more slowly than skeletal muscle, they can maintain these contractions for a longer period of time.

Cardiac Muscle. A third type of muscle is cardiac muscle. As its name implies, it is found only in the heart and it is responsible for the strong, regular contractions known as the heartbeat. Like smooth muscle, cardiac muscle is involuntary, but unlike any other type of muscle, cardiac muscle contracts independently of any nerve supply. This means that cardiac muscle contracts and relaxes automatically according to its own, built-in rhythm. Cardiac muscle is amazingly strong and resilient, given that it has to beat continuously over an individual's lifetime. Unlike skeletal muscle, which requires regular periods of rest, cardiac muscle neither requires—nor is allowed—to take a rest.

WHY MUSCLES CONTRACT

The actual contraction or shortening of any muscle occurs because a muscle contains two proteins, actin and myosin, that form long threads called filaments. These filaments are arranged in parallel stacks, like overlapping decks of cards. When a muscle gets a nerve signal to contract or shorten, the actin filaments physically slide over the myosin filaments and overlap them. The more they overlap, the more the muscle shortens. When the filaments slide back again, the muscle relaxes.

The human body has more than 600 muscles, and in the adult male, muscles make up about 40 percent of the total body mass. Although skeletal muscle fiber cannot replace itself by cell division after an organism is born, it can increase in size through exercise, which is important for healthy, strong muscles. Regular physical activity creates an increase in the number of blood vessels, which means the muscle receives more nutrients and oxygen. Muscles that are not used for a long period of time undergo a wasting process called atrophy.

[*See also* **Heart**]

Mutation

A mutation is any change in the genetic structure of a living cell. When that cell divides, the mutation is transmitted to the new cell and in the case of an organism, may result in offspring that look or behave different than its parents. Mutations are random events that can have a variety of causes.

The word mutation is credited to the Dutch botanist Hugo Marie De Vries (1848–1935), who observed random changes that suddenly appeared in the flowers he studied. De Vries named the changes mutations after the Latin word *mutare,* which means to change. De Vries had independently discovered Mendel's laws of inheritance, and when he added his theory of mutations (which meant there could be sudden "jumps" in evolution) to Mendel's laws, he was able to provide Darwin's theory of evolution with its missing mechanisms of change. We now know that mutations occur constantly in nature and that they are essential to life. Mutations are the source of variation in all living things, and without them populations would never evolve.

Mutations occur at the cellular level. Cells carry their codes for inherited characteristics in threadlike structures called chromosomes which,

A mutation caused this frog to be born with six legs instead of just four. (Reproduced by permission of JLM Visuals.)

in turn, carry genes that consist of a substance called deoxyribonucleic acid, or DNA. The coded information contained in the DNA determines the characteristics of living things. DNA looks like a spiral staircase or ladder, with the stair steps or ladder rungs determining the code for a particular organism. Any change or break in this DNA ladder alters the genetic code and results in a change in the next generation of the organism, whether it is a bacterium or a human being.

[*See also* **Chromosome; Genetic Disorders; Genetics**]

Natural Selection

Natural selection is the process of survival and reproduction of organisms that are best suited to their environment. It is a unifying idea for all of biology, and it also explains how the theory of evolution (the process by which living things change over generations) actually works.

The theory of evolution was first suggested by the English naturalist Charles Robert Darwin (1809–1882) in 1858. As a major theory, it attempted to account for the amazing diversity in the living world and to explain how present-day organisms came to be. It stated that all life progressed from simple to more complex organisms, and that gradual genetic changes occurred over a long period of time. Darwin's idea of natural selection was the key to explaining how this is accomplished. Natural selection has been described as the mechanism of evolution or even as the cause of evolution. The end result of natural selection is that organisms are able to adapt to their environment and change over time.

Natural selection has been described as the "survival of the fittest" because it is an unforgiving natural process that weeds out those traits that are less fit. This was explained by Darwin himself who said that natural selection rested on a few obvious facts. The first of these is that in the natural world, reproduction has the potential to produce more individual organisms than can survive. For example, a single leopard frog has the capability to produce 3,000 frogs a year. Since every species' living space or habitat has limited resources, population increases result in competition between individuals. Darwin said that at any one time, each organism is competing with others as well as members of its own species for such things as food, shelter, and members of the opposite sex with

CHARLES ROBERT DARWIN

English naturalist Charles Darwin (1809–1882) is without doubt one of the greatest and most influential life scientists who ever lived. His theory of evolution became the dominant concept for all of biology. Darwin's explanation that individual species can change or adapt over time, that humans have evolved from earlier "less human" forms, and that ultimately all life on Earth is connected and passed from simple to more complex organisms, is the "big idea" of biology.

Born in Shrewsbury, England, to a well-to-do family, the young Darwin had a physician father, and his two grandfathers were both wealthy and influential individuals. Expected to become a doctor like his father, Darwin could not tolerate watching surgery while in medical school, so his father suggested a career in the church. When the young man proved unwilling to pursue that profession, his father lost hope and declared he would grow up as a disgrace to the family. Throughout all this, however, the young man was starting to cultivate an interest in the natural world, and he eventually made natural history his hobby. His first real exposure to science was a field trip he took that was led by the English geologist (a person specializing in the origin, history, and structure of Earth) Adam Sedgwick (1785–1873). It was Sedgwick who recognized that there was something special about Darwin. This also enabled him to meet an English botany (the study of plants) professor, John Stevens Henslow (1796–1861), who helped Darwin obtain a position on a scientific expedition that would be making a five-year voyage to South America and the South Pacific Islands. Darwin's father eventually allowed him to go, and the twenty-two-year-old Darwin became an unpaid naturalist on the government survey ship, the H.M.S. *Beagle,* as it set sail in December 27, 1831. This grand voyage by sea not only transformed Darwin into a real naturalist (what is now called a biologist), but it proved to be one of the most important scientific voyages ever undertaken.

Darwin's job on the trip was to make geological and biological observations, keep records, and collect specimens. With each day, he learned a little more about the incredible variety in the natural world, and after awhile, he began to question why he found species that were closely related but still had noticeably different characteristics. Four years into his journey, Darwin landed on the Galapagos Islands far off the coast of Ecuador. There he noticed that there were about fourteen different types of finches on these different islands, with each bird apparently having adapted perfectly to its particular island environment. He also found that the natives could tell just by looking which island a giant tortoise had come from because of its distinctive features. All the while, Darwin searched for a pattern of meaning in this, and after some thought, he began to realize that it might be that species could actually change. It would make very good sense if one type of land finch colonized these islands and then each adapted, or changed, slightly to better fit its particular island. However, Darwin could not explain how this might occur, and he eventually returned home with no real answers.

After publishing his very popular book, *A Naturalist's Voyage on the Beagle,* Darwin began to seriously work out an explanation for the ideas he was considering. Those ideas were influenced not only by the writings of his good friend, the Scottish geologist Charles Lyell (1797–1875), but by those of the English economist, Thomas Robert Malthus (1766–1834). Malthus had written that a population always grows faster than does its food supply, and when Darwin applied this notion to the natural world, he realized that it might explain how species change. Since each individual is slightly different from every other, those that possess a certain trait that gives them some sort of advantage in competing for food would have a better chance of surviving and passing that trait on. A new species, therefore, is developed that is better able to survive in its environment.

In 1844, Darwin started a book that would explain his theory, but by 1858 he still did not have it completed. So when the English naturalist Alfred Russel Wallace (1823–1913) sent Darwin a draft of his own paper on this very same subject, the two decided to issue their papers together. Darwin, however, went on to elaborate much more fully in his 1859 book, *On the Origin of Species by Means of Natural Selection.* Although his book was immediately controversial, most of the scientific community was persuaded, and every copy sold out on its first day. With the post-1900 discovery of Austrian botanist Gregor Mendel's laws of inheritance (characteristics are not inherited in a random way but instead follow predictable, mathematical patterns), and the later discovery that genes are the basic units of heredity, Darwin's theory of evolution at last had a mechanism that explained exactly how it could take place.

Although Darwin lived a fairly long life, he was always troubled by a variety of physical problems, and while it was thought by many that he was probably a hypochondriac (someone who believes that they are always ill), many now think that he ruined his health during the *Beagle* trip by contracting a tropical disease. No book has been as important or as controversial as Darwin's, given that it goes against the Biblical view of creation. Although some today still disagree with its conclusions and implications, most life scientists agree that his theory remains the only viable scientific explanation for the amazing variety and diversity of life on Earth.

whom to mate. The organism's habitat or environment is the key to natural selection since it is the standard against which an individual's "fitness" is measured. Simply, if one organism possesses traits that make it more fit for its particular environment than does another (whether it is a longer neck, different color, larger antlers, or aggressive temperament), it is more likely that the "fitter" one will have a better chance to survive, to reproduce, and to pass on those advantageous traits. Since the environment is the measure, natural selection becomes an essentially random process. This means that nature has no master plan to favor one trait over another. Rather, a variety of traits usually exist in a given population, and whichever one happens to give its owner an edge over others becomes the trait that the environment favors.

GENETIC VARIETY

This leads to the second fact stated by Darwin: that close examination of any population of the same organisms reveals that not all individuals are exactly alike. In every group of like animals, there is always variety in both form and function. Simply, no two animals of the same species are identical because of the factor Darwin called "genetic variety." Genetic variety states that the individuals making up a population have inherited characteristics that make them slightly different from one another. This becomes obvious when we realize that if each individual were identical genetically, then it would make no difference which one survived to reproduce. But when genetic differences do exist, who gets to pass on what trait makes a very big difference.

Genetic variety comes about in two ways. The first is the physical result of sexual reproduction in which a unique individual is created who possesses a mixture of genes from both parents. This is called genetic recombination. The other way that genetic variety occurs is by mutations or accidental changes in a gene. A mutation is not necessarily something bad, and sometimes a chance change in a gene can result in a trait that gives an individual an advantage over others.

Finally, natural selection is always tied to reproduction, since it does no good for an organism to live a very long time if it does not reproduce. The way natural selection works is that

English naturalist Charles Darwin is considered the father of the theory of evolution by natural selection. (Photograph courtesy of The Library of Congress.)

those best suited to their environment (the fittest) survive better and get to produce more offspring, thus passing on their genes to future generations. The end result of natural selection is a process called "adaptation." Through natural selection, which favors organisms that fit their environment best and which weeds out those badly fitted, living things become better suited, fit, or "adapted" to their local environment or habitat. As this process continues over millions of years, new species evolve which are better adapted to their habitat or ecological niche (a specific job, or role, in a community that relates to feeding).

Although the theory of natural selection is popularly identified with Darwin, his contemporary, the English naturalist Alfred Russel Wallace (1823–1913), came up with the same theory in the same year. Darwin was at first amazed when he received Wallace's not-yet-published ideas in 1858, but the two men became allies and published their ideas together in a scientific journal that year. Later, Darwin went on to produce a fuller and more complete theory with his book, *On the Origin of Species,* and as a result received much more recognition than Wallace.

[*See also* **Evolution; Evolution, Evidence of; Evolutionary Theory; Human Evolution**]

Nervous System

The nervous system is a network of nerve cells that allows an animal to collect, process, and respond to information. As an internal communications system, the nervous system enables an animal to react and adjust to changes in its environment. Almost all animals have some type of nervous system, but the human nervous system allows us to speak, solve problems, and have creative ideas—activities that make humans different from all other animals.

Animals have a nervous system but plants do not. Plants react and respond to changes in their environment by slowly altering their growth patterns using different hormones. When a plant inclines itself toward a light source or drops its leaves, it is doing so on command from a naturally occurring chemical called a hormone that it produces as a response to something outside itself. Such a slow system of internal communication could only be practical for an organism that can make its own food and does not move. For animals, however, their very existence and reproductive ability often rests on being able to react immediately to something in their environment. Often they are either trying to catch something to eat or they are trying to escape being caught. Movement is, therefore, essential to animals, and their nervous systems must always allow them to act appropriately, efficiently, and most important of all, rapidly.

EVOLUTIONARY DEVELOPMENT OF THE NERVOUS SYSTEM

Even the simplest multicellular organism has to constantly gather and analyze information about its environment if it is to maintain its inner balance (known as homeostasis) and survive in a constantly changing habitat. The single-celled amoeba does not have a real nervous system, but it still responds appropriately to stimuli like light or food. However, more complex organisms need to do more than simply react to stimuli and therefore need a more complex communications system. Probably the simplest nervous system is the one used by the class of animals called Scyphozoa (phylum Cnidaria), better known as jellyfish, hydra, and sea anemones. These animals use a system described as a nerve net that directly connects the receiving cell to the cell that does the responding. Flatworms are more complex than jellyfish and concentrate both their receiving and sending sensors in their forward end (like a head). This means that their front part is the first to meet the stimuli, and this permits them to react more rapidly. Flatworms also have bilateral symmetry (a body that is basically the same on both halves). Bilateral symmetry includes a typical vertebrate system with pairing of nerves down either side of a central column.

Continuing up the ladder past the jellyfish and the flatworm, the beginnings of the vertebrate (an animal with a backbone) model can be seen inside an earthworm or a grasshopper. Besides having bilateral symmetry, they also demonstrate what is called segmentation. They have identical segments (each of which has a pair of nerves) linked together and connected to a central organ that could be called a primitive brain. Once this model was established and seemed to work well, it eventually evolved into the sophisticated nervous system that is basic to humans and all other animals with backbones. The backbone or the vertebral column evolved into its present form because it proved, first of all, to be an ideal up-and-down framework to support the body and make it both strong and flexible. Its hollow center was eventually expanded and modified so that it could hold a spinal cord connected to a brain. Finally, the skull that held the brain then slowly developed a range of sensory equipment (ears, eyes, nose) that gathered information about the outside world and fed it into the brain.

Opposite: The brain and the spinal cord are the two major components of the central nervous system, which is considered the command center of the body. (Illustration of Kopp Illustration, Inc.)

THE CENTRAL NERVOUS SYSTEMS

Today, the nervous system of all vertebrates can be divided into two main parts: the central nervous system and the peripheral nervous sys-

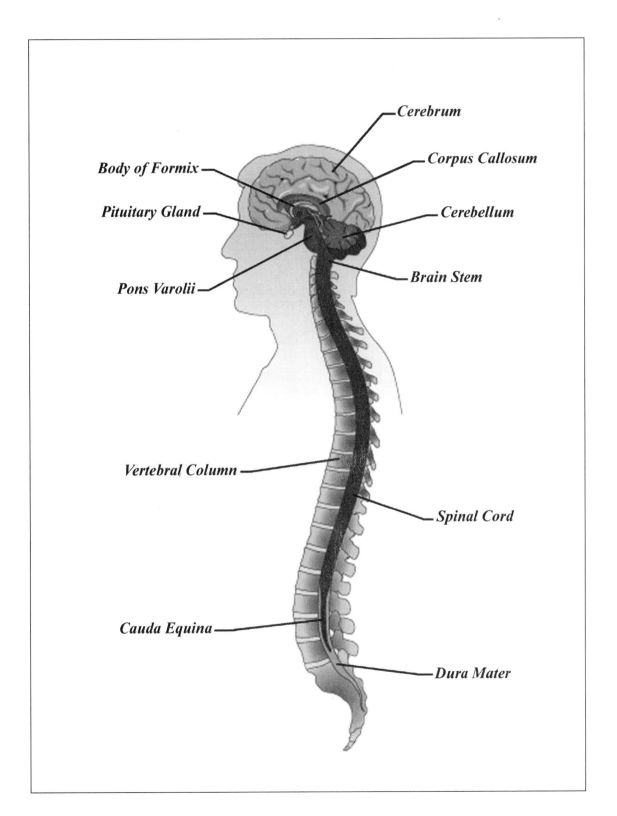

Cerebrum

Corpus Callosum

Body of Formix

Pituitary Gland

Cerebellum

Brain Stem

Pons Varolii

Vertebral Column

Spinal Cord

Cauda Equina

Dura Mater

SANTIAGO RAMON Y CAJAL

Spanish histologist (a person specializing in the study of tissues and organs) Santiago Ramon y Cajal (1852-1934) laid the foundations of modern neurology, which is the scientific study of the nervous system. He established that the nervous system is made up of independent units of nerve cells called neurons, and also made important discoveries relating to the transmission of nerve impulses and the cellular structures of the brain.

Santiago Ramon y Cajal was born in the remote village of Petilla de Aragon, Spain. His father was a self-trained country doctor who wanted his son to have a real medical education. Therefore, the family moved to the university city of Zaragoza, and young Santiago studied medicine there. Santiago was a rebellious youth who was more interested in drawing than studying, and his father finally forced him to work for a barber and a shoemaker as a way of disciplining him and making him appreciate school. This apparently worked, and he earned his medical degree from the University of Zaragoza in 1873. He then joined Spain's army medical service and served as army surgeon in Cuba for a year. However, while in Cuba he caught malaria and returned to Spain to recuperate. Going back to school, he earned his doctorate in medicine in 1879 and began a teaching career. Preferring to do research and to teach rather than practice medicine, he turned to the field of anatomy, which had always been his favorite subject. Anatomy studies the structure of living things and figures out how the different parts of an organism are shaped and how they fit together. His favorite branch of anatomy was histology.

Working with an old, abandoned microscope he found at the University of Zaragoza, Ramon y Cajal eventually turned toward the study of the most

tem. The central nervous system consists of the brain and spinal cord and acts as network central, or the "main switchboard" for the entire system. The cord itself lies within and is protected by a central canal that is made up of stacks of bony vertebrae. Like an electrical cable, it has pathways that lead to and from the brain, and its nerve cells are lined up in a columnar way. Inside the backbone, the spinal column is bathed in cerebrospinal fluid that circulates around both it and the brain and acts as a liquid shock absorber. At the top of the column sits the brain—the system's real control center. It is actually a continuation of the spinal cord, which extends upwards and expands into segments called ventricles and aqueducts. The brain itself has three parts: the hindbrain, the midbrain, and the forebrain. The hindbrain occupies the rear

complex tissues in the body, which are those of the nervous system. These can only be studied under the microscope if they are stained (dyed), and Ramon y Cajal was able to improve upon some of the stains in use at this time. When he began his research on the nervous system, little or nothing was known about what might be called the path that a nervous impulse takes. Most scientists thought it traveled over something like a connected grid, or network of wires, but no one really knew for sure since they had never been able to examine nervous tissue closely. Using his own improved stain techniques on brain tissue, Ramon y Cajal was able, by 1889, to demonstrate that the nervous system was by far much more complex than anyone had imagined. He then went on to show that the neuron, or nerve cell, was the basic unit of the nervous system, and that it was very different from the other ordinary cells of the body. Ramon y Cajal then offered a controversial "neuron theory" that few accepted. He stated that the nervous system consists of a network of separate nerve fibers whose ends never actually "touch" or are not actually connected with the surrounding nerve cells. Today it is known that neurons are indeed not "hard-wired" together but instead have a synapse or space between them, across which the electrical charge or nervous impulse "jumps." He also stated that nerve impulses travel in only one direction, and that the brain's neurons had different structural patterns in different areas, suggesting that one part might have a different job than another. We now call this "localization," meaning that a certain part of the brain controls memory and another intelligence. Ramon y Cajal also conducted important research on the tissues of the inner ear and eye. Overall, modern neurology really began with his work, since he established the correct role played by the neuron and the nervous impulse.

of the skull and connects with the cord. It contains the medulla oblongata and the cerebellum. The medulla controls the body's involuntary processes like breathing and the heartbeat. The cerebellum coordinates the body's many muscles and allows it to move properly. The midbrain is sort of a coordinating center for information collected by sight and hearing, and it also relays information to higher centers of the brain. The forebrain contains the higher brain centers such as the cerebrum. The cerebrum is the part of the brain involved in voluntary actions and with functions related to memory, learning, and sensing things. In humans, the cerebrum is larger than the hindbrain and midbrain put together, and is the place that governs thinking, reasoning, and the use of complex language.

THE PERIPHERAL NERVOUS SYSTEM

The peripheral nervous system can be described as a branching network of nerves that carry signals to and from the central nervous system. This system runs throughout the body of a vertebrate and has two different types of nerves: afferent nerves (or sensory neurons) bring input into the central nervous system, while efferent nerves (or motor neurons) take output away and communicate it to the muscles and glands. Some nerves in the peripheral nervous system can be controlled voluntarily, such as those that allow us to walk or run, but those that are not under our control (like the production of saliva or our beating heart) are considered a division of the peripheral nervous system called the autonomic nervous system. This system regulates functions like digestion that we cannot control. These nerves keep the body running smoothly by automatically adjusting its many systems.

NEURONS

The entire system works as it does because it is based on the neuron or nerve cell, the fundamental unit of communication in all nervous systems. Neurons never act alone. Rather, they transmit impulses to one another in the form of electrical signals and link together the entire nervous system. In many ways, neurons are the body's electrical highway. The neuron is a specialized cell and consists of three parts: the soma, the dendrites, and the axons. The soma is the cell body with its cytoplasm and nucleus. Dendrites and axons are hairlike arms or branches that extend from the body and channel information in one direction. Dendrites form the input part of the system and carry information toward the cell body. Axons are for output and therefore carry information away from the cell body. A typical neuron has several dendrites and one axon. Finally, neurons pass impulses to one another in a one-way direction across a space called a synapse. When a nerve impulse reaches the end of an axon and arrives at the synapse, a transmitter substance is released from the axon into the synapse. This chemical neurotransmitter goes across and binds to a receptor in the adjoining dendrite and triggers an impulse in it. The brain of the average adult contains about 1,000,000,000 neurons. Since the functioning of the entire system is dependent on the precise and proper functioning of the different types of neurons, anything that interrupts or disturbs that synaptic function can cause a problem with the organism. Today, scientists know of many genetic and infectious diseases that can severely interfere with this function. Many drugs have also been developed that can positively or negatively affect it.

[*See also* **Brain; Muscular System**]

Niche

The term niche refers to the particular job, or function, that a living thing plays in the particular place it lives. Also called an ecological niche, this concept refers to the precise way in which an organism fits into its environment. A niche includes all the factors that are important to the organism's existence. No two species can occupy the same niche.

In order to study a niche, life scientists must understand all of the factors that are important to a living thing's existence. These factors include: an organism's: diet; energy, light, and moisture requirements, ideal temperature; ideal habitat; and ideal reproduction conditions. Therefore, the word function is key to understanding the idea of an ecological niche.

The idea of niche as a function has also been described as the job a living thing has—and as with any job—it can be very specialized or very general. Some organisms have very broad niches, meaning that they are fairly flexible in terms of the type of things they eat and temperature they tolerate. In other words, their overall living needs can be met in a less specialized environment. The opossum and the raccoon are examples of animals with very broad niches since they eat a wide variety of plants and animals and adjust well to different climates. However, many species play a narrow or specialized role and therefore have a narrow niche. Two examples are the giant panda of central China and the koala of Australia. The panda lives in the bamboo forests and eats only one type of bamboo, while the koala can only live where certain species of eucalyptus trees grow and survives by eating the leaves. Other factors influence a niche, and in the case of the koala, its existence becomes more fragile since it must live in a warm climate and it does not produce its young in great numbers. Not surprisingly, animals with the broader niche survive fairly easily and thrive in certain areas, while those animals with the more specialized niche, like the koala and the giant panda, are more likely to become endangered species as their habitats are destroyed.

Some life scientists say that a niche can be understood primarily in terms of competition, while others say it has more to do with one species being best fitted for a certain role. As early as the 1930s, life scientists developed the principle called competitive exclusion. One of the pioneers arguing for the primary importance of competition was the Russian microbiologist Georgil F. Gauze (1910–), who conducted tests on different species of protozoa (one-celled organisms). Gauze successfully raised different species of protozoa, each in its own environment. He then put them together and discovered that one died out completely while the other thrived. In another test, he found that although the two species survived, neither did so in great numbers, and each occupied a different territory (or a separate part of the test tube). From this work came Gauze's now-accepted principle that although two species may occupy the same habitat, they never share the same niche. A common example of such a phenomenon is that of the woodpecker and nuthatch. Both are birds that eat grubs (insects) that they find under tree bark. However, despite the fact that they are both after the same meal, neither occupies the niche of the other since woodpeckers start at the bottom of a tree and work their way up, while nuthatches begin at the top and work down.

Beyond an organism's feeding habits, there are scores of other factors that go into describing an organism's niche. For example, the simple earthworm plays a key role in its habitat as a consumer of dead organic matter and as food for other animals (such as birds). It is a host for certain parasites, and its burrowing has a beneficial plowing effect on the soil, loosening it up and allowing air to circulate. As the example of the earthworm shows, understanding an organism's niche also allows scientists to better understand both the organism and the environment it lives in.

Nitrogen Cycle

The nitrogen cycle describes the stages in which the important gas nitrogen is converted and circulated from the nonliving world to the living world and back again. There are five main steps in the nitrogen cycle, four of which are carried out by bacteria. Since nitrogen cannot be used by the cells of living things until it has been converted by bacteria into a useful form, these nitrogen-fixing bacteria not only play a key role in the nitrogen cycle but are in fact essential to all life on Earth.

All plants and animals need a certain amount of nitrogen in their systems in order to live. Nitrogen is an important component of amino acids, which are the building blocks of life, and nucleic acids, which make up genetic material. Although nitrogen is a free gas (it remains a gas under normal atmospheric conditions) that makes up nearly 80 percent of Earth's atmosphere, only bacteria can use it in this form. This means that for every plant and animal on Earth, the abundant nitrogen in the atmosphere is entirely useless unless it is somehow combined with hydrogen or oxygen. Bacteria that produce compounds of nitrogen, called nitrates, work to form this necessary combination. Most nitrates are produced by bacteria in the soil and are therefore absorbed by plants. Animals obtain the nitrates either by eating plants or other animals that eat plants. When plants and animals die, the nitrates in their systems are returned to the atmosphere and back to the beginning of the nitrogen cycle.

THE FIVE STEPS OF THE NITROGEN CYCLE

The nitrogen cycle has five main steps: fixation, nitrification, assimilation, ammonification, and denitrification. All except the third step (assimilation) are carried out by bacteria.

Nitrogen Fixation. The first step, nitrogen fixation, is probably the most important since it frees what was previously unusable nitrogen. Only a few types of bacteria are able to break the chemical bonds that hold the paired atoms in nitrogen gas together. Nitrogen-fixing bacteria that live in the soil have an enzyme called nitrogenase that they use to break that bond or "fix" and capture nitrogen gas from the atmosphere and convert it to useable ammonia. Blue-green algae, called cyanobacteria, is one example of bacteria that can change atmospheric nitrogen to ammonia.

Nitrification. The second step in the cycle, called nitrification, occurs when the converted nitrogen is combined with oyxgen and hydrogen. This combination results in the production of nitrogen compounds, like nitrates,

A flowchart showing all five steps of the nitrogen cycle beginning with atmospheric fixation. (Illustration by Hans & Cassidy. Courtesy of Gale Research.)

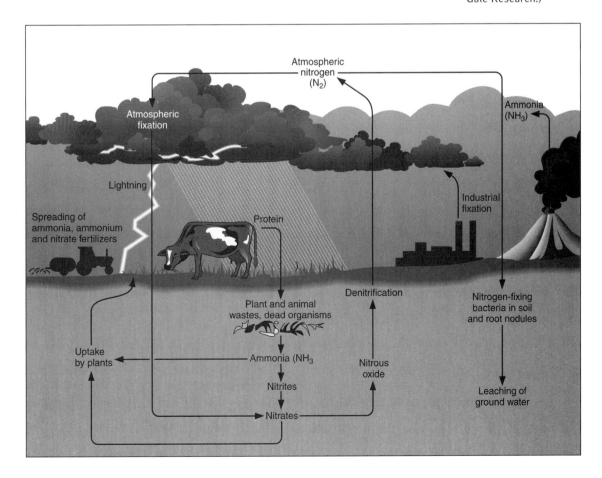

that other organisms can use. During the third stage, assimilation, plants take up these nitrates with their roots and use the nitrates to build proteins and nucleic acids. In other words, the plants use the nitrates to grow. Animals, like plants, also need proteins to function and grow, and proteins are composed of amino acids that contain nitrogen atoms.

Ammonification. The fourth stage in the nitrogen cycle is called ammonification. This is the process in which these nitrogen compounds are returned to the soil and reconverted to ammonia by other bacteria. Ammonification occurs in two ways. The first is when a plant or animal dies and their organic matter is acted upon by decomposer bacteria. These bacteria play an important role because the nitrogen trapped in dead animals and plants would be useless and wasted were it not converted. A second way for ammonification to occur is when animals leave their urine and feces on the ground. Almost immediately by the same decomposer bacteria work to release the nitrogen found in these waste products.

Denitrification. The last step in the nitrogen cycle is called denitrification and involves the reduction of ammonia to nitrogen gas. Again, this process is carried out by a certain type of bacteria whose actions eventually release nitrogen gas back into the atmosphere. The complete nitrogen cycle is, therefore, a process in which nitrogen from the atmosphere is captured and converted, passes through a number of organisms (including bacteria), and is finally returned to the atmosphere to enter the cycle again.

There is a more dramatic way that nitrogen is converted to a usable form. The atmosphere of Earth is roughly four parts nitrogen to one part oxygen, and these two gases never normally react with one another. However, during a thunderstorm when lightning strikes, the temperature and the pressure of the air is such that the two gases react chemically and form nitrous oxide gases. In the rainwater that accompanies the storm, nitrous oxide is dissolved and falls to Earth as nitric acid. This enters the nitrogen cycle when it soaks into the soil and forms nitrites and nitrates. However, far more nitrates are formed by bacteria in the soil than by atmospheric lightning flashes.

Nonvascular Plants

Nonvascular plants are plants that do not have any special internal pipelines or channels to carry water and nutrients. Instead, nonvascular plants absorb water and minerals directly through their leaflike scales. Nonvascular plants are usually found growing close to the ground in damp, moist places.

Nonvascular plants are made up of mosses, liverworts, and hornworts—all of which belong to the subdivision of plants called Bryophyta or bryophytes. They are described as nonvascular because they do not have an internal transport or circulatory system as do other plants. Bryophytes are also nonflowering plants, meaning that they reproduce without growing flowers. Lacking a system to move food and minerals, these plants are unable to grow very tall since they depend on direct contact with moisture. Bryophytes lack true leaves and do not have roots, using rhizoids instead. Rhizoids are slender, rootlike hairs that both anchor and absorb like roots.

After rhizoids perform this initial absorption, movement throughout the plant takes place by the processes of diffusion and active transport. Diffusion is the distribution of a substance away from an area where it is highly concentrated. In this way, water and nutrients move from cells that are full to cells that are empty. Diffusion uses no energy. Active transport, however, does require a plant to use energy. It is used when a plant needs to achieve the opposite of diffusion and must try to concentrate a substance in one place. A plant carries out active transport by using carrier protein molecules. These molecules actually carry the needed substance from one side of a cell membrane to the other. Because of this essentially primitive means of obtaining and moving water and nutrients, a humid, moist environment is essential to nonvascular plants.

[*See also* **Plant Anatomy; Plants**]

Nuclear Membrane

The nuclear membrane is the outermost part of a cell's nucleus that separates it from the cytoplasm. Also called the nuclear envelope, this double-membrane structure acts as a boundary for the nucleus, allowing it to keep its shape. It also allows controlled exchanges through its pores.

The nucleus is by far the most important structure in any cell, plant or animal, since it functions as the control center directing all of the cell's activities. The nucleus contains the chemical instructions deoxyribonucleic acid (DNA) needed to make a cell work properly. The nucleus is usually the largest separate structure in a cell and it typically has a round or oval shape. It keeps this shape because it has a double-layer membrane that keeps it separate from the rest of the cell's cytoplasm. Cytoplasm is the jelly-like contents of a cell that contains all of its other structures.

Besides acting as a boundary that keeps the nucleus together, the nuclear membrane also controls what passes between the nucleus and the cytoplasm. It carries out this regulating function by using nuclear pores that dot its surface. These pores are like a sieve (a strainer with certain size holes) and they allow small molecules in and out of the nucleus. They also selectively permit large molecules to pass through their openings.

The nuclear membrane also plays a key role during mitosis (my-TOH-sis), which occurs when a cell makes a copy of itself. During the later stages of mitosis, the nuclear membrane begins to break down, allowing the already-duplicated chromosomes to split into two groups. After each complete set of chromosomes moves to opposite ends of the cell, a nuclear membrane reforms around each group. Soon each new cell has a separate nucleus surrounded by its own nuclear membrane. This reforming of the nuclear membrane begins the completion phase of mitosis.

[*See also* **Cell; Membrane; Mitosis; Nucleus**]

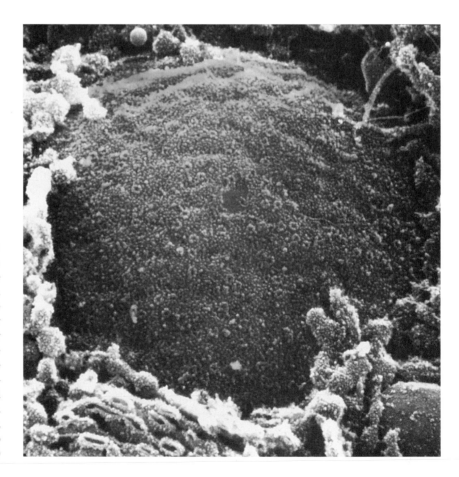

A high resolution scanning electron micrograph of the nuclear membrane. The nuclear membrane is the double layer with the granular appearance surrounding the chromosome-containing cell nucleus. (©Photographer, Science Source/Photo Researchers.)

Nucleic Acids

Nucleic acids are a group of organic compounds that carry genetic information. Nucleic acids are essential to life since they contain not only a cell's genetic information, but also instructions for carrying out cellular processes. Deoxyribonucleic acid (DNA) is the particular type of nucleic acid out of which genes are made, and genes are the bearers of hereditary traits from parents to offspring.

Nucleic acid was discovered in 1869 by the Swiss biochemist Johann Friedrich Miescher (1844–1895), who first found a sticky, clear chemical in the nucleus of cells. He named it nuclein, and although it later became known as nucleic acid, no one had any idea that it was in some way connected to heredity. The name nucleic acid itself indicates that these clear molecules were first found in the cell nucleus and that they have a mildly acidic character. By 1929, scientists had discovered that there were two types of nucleic acids. One of these contained the sugar ribose (and became known as ribonucleic acid or RNA) and the other contained the sugar deoxyribose (and was called deoxyribonucleic acid or DNA). By the 1930s, most geneticists agreed that the gene was crucial to heredity and was made of some sort of complex chemical, but no one thought it was made of nucleic acid because the acid did not seem to have a complicated enough structure to carry genetic information.

By 1950, however, nucleic acid had been established as the key factor in inheritance, yet in the fall of that year when the young American biochemist James Dewey Watson (1928–), traveled to Europe to study the chemistry of nucleic acids, no one knew how this chain of fairly simple molecules could contain all the information necessary to form a complex organism. In 1951 when Watson met the English biochemist Francis Harry Compton Crick (1916–) at the Cavendish Laboratory at Cambridge, England, the two began a collaboration with the goal of determining the structure of DNA (which they believed would then explain how DNA actually works).

In March 1953, the team of Watson and Crick announced that they had discovered the "double helix structure" of the DNA molecule and offered to science what was essentially an explanation of the chemical basis of life itself. Their theory was that the nucleic acid DNA was made up of two twisted strands that are held together by base pairs that make up the actual coded instructions. Each DNA base is, therefore, like a letter in the alphabet, and a sequence of bases can be thought of as forming a message. Nucleic acids were found not only to contain genetic information (DNA), but were also able to carry that information from genes in the nucleus to other structures in the cell. Thus, the building of proteins was found to be controlled by the group of nucleic acids known as ribonucleic acid (RNA). Geneticists eventually came to describe RNA according to its function in the cell, messenger RNA (mRNA) and transfer RNA (tRNA). Both types of RNA are essential for a DNA molecule to make a copy of itself (which in turn is how proteins are made). mRNA passes out of the nucleus and carries the message for making a protein. tRNA reads this message and transfers the right amino acid to where they form proteins. Watson and Crick's landmark discovery of the nature and function of nucleic acids provided the foundation for understanding the chemical basis of life.

[*See also* **Chromosome; DNA; Enzyme; Genetic Engineering; Genetics; RNA**]

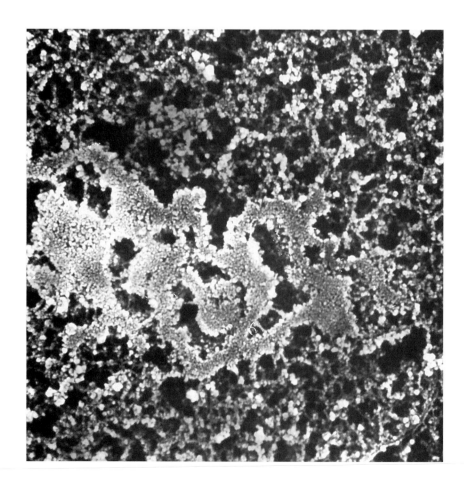

A high resolution scanning electron micrograph of the nucleolus of an eight-week-old human embryo. (©Photographer, Science Source/Photo Researchers.)

Nucleolus

The nucleolus is the part of the nucleus (the cell's control center) of a cell that helps produce ribosomes. Ribosomes are those parts of a cell that help make proteins. The nucleolus is easily recognized as a dark, dense area near the center of the nucleus.

Most cells have only one nucleolus, although some have two or more. When a cell nucleus is stained in order to observe it better under a microscope, the nucleolus is always seen as a dark-stained body. Its shape is usually irregular, probably because it is not walled off from the rest of the nucleus by any type of membrane. It looks like a mass rather than something that is sharply defined.

The nucleolus has an important job, however, in that it assembles ribosomes. After the nucleolus assembles them, the ribosomes leave the nucleus through its membrane's pores and enter the cell's cytoplasm (the jelly-like contents of a cell). Here they go to work making proteins. Since nucleoli are indirectly involved in making proteins, they perform a key function in the cell. Rapidly growing cells require a great deal of protein, which means that they must also have a lot of ribosomes. The name nucleolus means "little nucleus," and the nucleolus does resemble a small nucleus within a large nucleus.

[*See also* **Cell; Nucleus**]

Nucleus

The nucleus is the control center that directs the activities of the cell. Most important is its control of cell reproduction and the construction of materials like proteins. The nucleus also functions as the cell's main repository of genetic information in the form of deoxyribonucleic acid (DNA).

A eukaryotic cell is a cell that contains a separate nucleus. Plants, animals, fungi, and some forms of single-celled life are eukaryotic or eukaryotes. Those living things, like bacteria, whose cells do not have a distinct nucleus are called prokaryotes. A eukaryotic cell contains many structures called organelles, each of which has a separate function. The most important organelle in a cell is its nucleus (named after a Latin word meaning "kernel" because it looks like a seed in the center of a fruit). If a cell can be described as a miniature factory in which conditions are carefully regulated, then the nucleus in the cell is the factory's main office or control center. The nucleus is a very busy place as it simultaneously controls many cellular activities, responds to changes in its environment, and helps make ribonucleic acid (RNA). As the heart of every cell, the nucleus is easily seen when a cell is observed with a microscope. Because of its large size, the nucleus is by far the outstanding feature of every cell. It usually appears as a rounded structure near the center of the cell. Besides its size, what makes the nucleus so distinct is that it is surrounded by a double membrane called the nuclear envelope. This keeps the nucleus separate from the rest of the cell's living material called the cytoplasm. This envelope has many tiny openings known as pores that allow certain substances to pass in and out of the nucleus to the rest of the cell.

Inside the nucleus are two important structures: threadlike structures called chromosomes and small, round structures called nucleoli. The chromosomes contain the cell's genetic material called DNA, which contains all the instructions needed to make a cell work. DNA also contains a cell's genes, which are the basic units of heredity. The number of chromosomes a nucleus contains will change from species to species (humans have forty-six chromosomes). Also found in the nucleus are one or more small, round nucleoli (singular, nucleolus) which help the cell make ribosomes. Ribosomes are organelles that play an important role in the manufacture of proteins. The nucleolus sends RNA to the ribosomes, which use amino acids to make proteins. Ribosomes are outside the nucleus. The nucleus of a cell

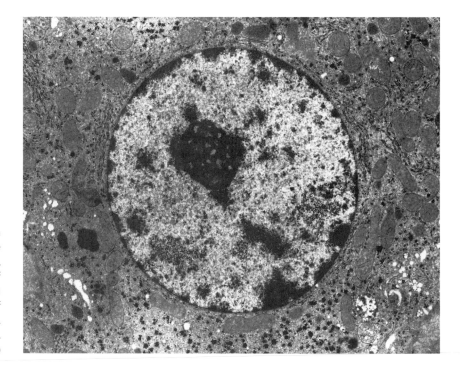

The nucleus and perinuclear area in a rat's liver cell. The nucleus is the dark area in the middle of the cell. (Reproduced by permission of Phototake. Photograph by Dr. Dennis Kunkel.)

is therefore the control center of the cell because it directs the cell's activities by controlling the synthesis of production of proteins. It is with these chemicals called proteins that the nucleus actually runs the cell. Proteins are also a key ingredient in the material out of which cells, are made.

[*See also* **Cell; Cell Division; Chromosome; DNA; Protein**]

Nutrition

Nutrition is the process by which an organism obtains and uses raw materials from its environment in order to stay alive. Autotrophs, like plants, are able to make their own nutrients, while heterotrophs, like animals, must ingest or eat them. Nutritional needs of plants and animals are different. These needs may vary considerably for animals according to their age, sex, level of activity, and reproductive status.

Nutrients are any substances that are taken in by a living thing in order to survive or to sustain its life. All living things must take in nutrients from their environment in order to grow and repair themselves and to provide the energy they need. This process of taking in nutrients and using them is called nutrition.

AUTOTROPHS

Plants and animals not only have different nutritional requirements, but they obtain their nutrients in different ways. Plants are called autotrophs because they produce many of their own nutrition requirements by photosynthesis. Photosynthesis is the chemical conversion of the Sun's light into simple food chemicals. A plant makes its own nutrients from inorganic materials like nitrogen (gas), carbon dioxide (gas), and sunlight. The basic elements that a plant needs are mostly in the soil in which they grow. Elements required in large amounts are called macronutrients. These macronutrients are carbon, oxygen, hydrogen, nitrogen, phosphorous, potassium, calcium, and magnesium. Plants also require smaller amounts of micronutrients such as chlorine, iron, boron, manganese, zinc, copper, molybdenum, and nickel. Plants, or autotrophs, are considered the "primary producers" in the food chain (the series of stages energy goes through in the form of food) because all other life depends on them directly or indirectly.

HETEROTROPHS

Heterotrophs include humans and other animals. Heterotrophs cannot make organic material the way plants can. Instead, they must obtain their nutrients from the food they eat. Heterotrophs are divided into herbivores that eat plants, carnivores that eat other animals, and omnivores who eat both plants and animals. Humans are omnivores and must obtain their nutrients from plants and animals. Like plants, the nutrients humans take in are also divided into macronutrients and micronutrients.

Macronutrients are substances humans need in large amounts, like carbohydrates, proteins, and fats. All three of these essential nutrients are large, complex molecules that must be digested, or broken down, into smaller units so they can be absorbed or taken into the body's cells. Micronutrients are substances like vitamins and minerals that people need in much smaller amounts. However, just because micronutrients are not needed in great quantity does not mean that they are not essential to the proper functioning of the body. Vitamins are especially important because they help enzymes (proteins that control the rate of chemical changes) regulate chemical reactions and are not generally stored by the body. This means that vitamins must be obtained by one's diet.

The food pyramid developed by the U.S. Department of Agriculture. The bottom level consists of cereal foods, the second of fruits and vegetables, the third of proteins, and the fourth of fats and oils. For a healthy, nutritious diet, one should eat more of the foods represented in the lower levels. (Illustration courtesy of Gale Research.)

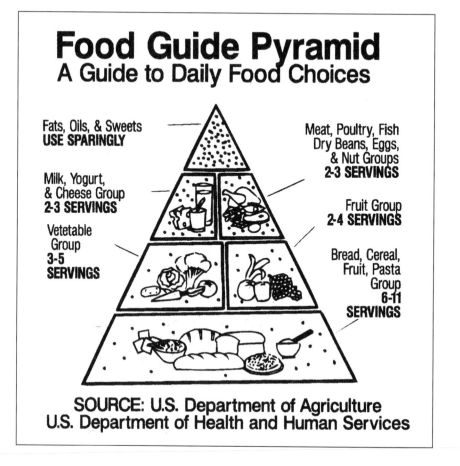

Food Guide Pyramid
A Guide to Daily Food Choices

Fats, Oils, & Sweets
USE SPARINGLY

Meat, Poultry, Fish
Dry Beans, Eggs,
& Nut Groups
2-3 SERVINGS

Milk, Yogurt,
& Cheese Group
2-3 SERVINGS

Fruit Group
2-4 SERVINGS

Vetetable
Group
**3-5
SERVINGS**

Bread, Cereal,
Fruit, Pasta
Group
**6-11
SERVINGS**

**SOURCE: U.S. Department of Agriculture
U.S. Department of Health and Human Services**

For all living things, food is a key to life since it contains the raw materials, or nutrients, needed to stay alive. Food performs three functions for animals. It acts as fuel for all of the various reactions that take place in the human body, and it provides energy to move about. Food also provides the organic raw materials for the body to make new structures and to repair others. Finally, food contains the nutrients that animals cannot make for themselves. A good example are the much-needed ten "essential amino acids" that the body cannot make on its own, but must be obtained through food.

MALNUTRITION

The condition known as malnutrition literally means "bad nutrition," and an organism is described as malnourished when it is not getting enough of one or more nutrients. The lack of enough carbohydrates and fats means that eventually, the body will start to consume its own proteins. This only happens when a person is actually starving, and it leads to a wasting away that eventually proves fatal.

VARYING NUTRITIONAL NEEDS

The absence of an important vitamin usually leads to a specific deficiency disease, and the lack of a needed mineral also can lead to serious conditions. For humans, a diet that supplies the proper amount of nutrients is said to be "balanced," but a person's nutritional needs vary greatly according to their stage in life. During infancy, a person's total energy requirements are the highest. Later in childhood, growth is rapid, and protein intake should be fairly high to develop new tissue. During adolescence, growth spurts also occur, which make impressive nutritional demands. In adulthood, growth has stopped and nutritional requirements are less, although they are still very important to maintaining good health.

A person's level of activity also affects their nutritional needs, and a pregnant woman or an athlete both have different nutritional needs compared to an inactive senior citizen. Although the nutrient amounts taken in depends on these and other variables, a general rule of thumb would be that carbohydrates should provide half of the total energy intake; fats should make up about one third; and protein should comprise the remainder.

[*See also* **Amino Acid; Carbohydrate; Lipids; Malnutrition; Protein; Vitamin**]

For Further Information

Books

Abbot, David, ed. *Biologists.* New York: Peter Bedrick Books, 1983.

Agosta, William. *Bombardier Beetles and Fever Trees.* Reading, Mass.: Addison-Wesley Publishing Co., 1996.

Alexander, Peter and others. *Silver, Burdett & Ginn Life Science.* Morristown, N.J.: Silver, Burdett & Ginn, 1987.

Alexander, R. McNeill, ed. *The Encyclopedia of Animal Biology.* New York: Facts on File, 1987.

Allen, Garland E. *Life Science in the Twentieth Century.* New York: Cambridge University Press, 1979.

Attenborough, David. *The Life of Birds.* Princeton, N.J.: Princeton University Press, 1998.

Attenborough, David. *The Private Life of Plants.* Princeton, N.J.: Princeton University Press, 1995.

Bailey, Jill. *Animal Life: Form and Function in the Animal Kingdom.* New York: Oxford University Press, 1994.

Bockus, H. William. *Life Science Careers.* Altadena, Calf.: Print Place, 1991.

Borell, Merriley. *The Biological Sciences in the Twentieth Century.* New York: Scribner, 1989.

Braun, Ernest. *Living Water.* Palo Alto, Calf.: American West Publishing Co., 1971.

Burnie, David. *Dictionary of Nature.* New York: Dorling Kindersley Inc., 1994.

Burton, Maurice, and Robert Burton, eds. *Marshall Cavendish International Wildlife Encyclopedia.* New York: Marshall Cavendish, 1989.

Coleman, William. *Biology in the Nineteenth Century.* New York: Cambridge University Press, 1977.

Conniff, Richard. *Spineless Wonders.* New York: Henry Holt & Co., 1996.

Corrick, James A. *Recent Revolutions in Biology.* New York: Franklin Watts, 1987.

Curry-Lindahl, Kai. *Wildlife of the Prairies and Plains.* New York: H. N. Abrams, 1981.

Darwin, Charles. *The Origin of Species.* New York: W.W. Norton & Company, Inc., 1975.

Davis, Joel. *Mapping the Code.* New York: John Wiley & Sons, 1990.

Diagram Group Staff. *Life Sciences on File.* New York: Facts on File, 1999.

Dodson, Bert, and Mahlon Hoagland. *The Way Life Works.* New York: Times Books, 1995.

Drlica, Karl. *Understanding DNA and Gene Cloning.* New York: John Wiley & Sons, 1997.

For Further Information

Edwards, Gabrielle I. *Biology the Easy Way.* New York: Barron's, 1990.

Evans, Howard Ensign. *Pioneer Naturalists.* New York: Henry Holt & Sons, 1993.

Farrington, Benjamin. *What Darwin Really Said.* New York: Schocken Books, 1982.

Finlayson, Max, and Michael Moser, eds. *Wetlands.* New York: Facts on File, 1991.

Goodwin, Brian C. *How the Leopard Changed Its Spots: The Evolution of Complexity.* New York: Simon & Schuster, 1996.

Gould, Stephen Jay, ed. *The Book of Life.* New York: W.W. Norton & Company, Inc., 1993.

Greulach, Victor A., and Vincent J. Chiapetta. *Biology: The Science of Life.* Morristown, N.J.: General Learning Press, 1977.

Grolier World Encyclopedia of Endangered Species. 10 vols. Danbury, Conn.: Grolier Educational Corp., 1993.

Gutnik, Martin J. *The Science of Classification: Finding Order Among Living and Nonliving Objects.* New York: Franklin Watts, 1980.

Hall, David O., and K.K. Rao. *Photosynthesis.* New York: Cambridge University Press, 1999.

Hare, Tony. *Animal Fact-File: Head-to-Tail Profiles of More than 100 Mammals.* New York: Facts on File, 1999.

Hare, Tony, ed. *Habitats.* Upper Saddle River, N.J.: Prentice Hall, 1994.

Hawley, R. Scott, and Catherine A. Mori. *The Human Genome: A User's Guide.* San Diego, Calf.: Academic Press, 1999.

Huxley, Anthony Julian. *Green Inheritance.* New York: Four Walls Eight Windows, 1992.

Jacob, François. *Of Flies, Mice, and Men.* Cambridge, Mass.: Harvard University Press, 1998.

Jacobs, Marius. *The Tropical Rain Forest.* New York: Springer-Verlag, 1990.

Johanson, Donald, and Blake Edgar. *From Lucy to Language.* New York: Simon & Schuster, 1996.

Jones, Steve. *The Language of Genes.* New York: Doubleday, 1994.

Kapp, Ronald O. *How to Know Pollen and Spores.* Dubuque, Iowa: W. C. Brown, 1969.

Kordon, Claude. *The Language of the Cell.* New York: McGraw-Hill, 1993.

Lambert, David. *Dinosaur Data Book.* New York: Random House Value Publishing, Inc., 1998.

Leakey, Richard. *The Origin of Humankind.* New York: Basic Books, 1994.

Leakey, Richard, and Roger Lewin. *Origins Reconsidered.* New York: Doubleday, 1992.

Leonard, William H. *Biology: A Community Context.* Cincinnati, Ohio: South-Western Educational Pub., 1998.

Levine, Joseph S., and David Suzuki. *The Secret of Life: Redesigning the Living World.* Boston, Mass.: WGBH Boston, 1993.

Little, Charles E. *The Dying of the Trees.* New York: Viking, 1995.

Lovelock, James. *Healing Gaia.* New York: Harmony Books, 1991.

McGavin, George. *Bugs of the World.* New York: Facts on File, 1993.

McGowan, Chris. *Diatoms to Dinosaurs.* Washington, D.C.: Island Press/Shearwater Books, 1994.

McGowan, Chris. *The Raptor and the Lamb.* New York: Henry Holt & Co., 1997.

McGrath, Kimberley A. *World of Biology.* Detroit, Mich.: The Gale Group, 1999.

Magner, Lois N. *A History of the Life Sciences.* New York: Marcel Dekker, Inc., 1979.

Manning, Richard. *Grassland.* New York: Viking, 1995.

Margulis, Lynn. *Early Life.* Boston, Mass.: Science Books International, 1982.

Margulis, Lynn, and Karlene V. Schwartz. *Five Kingdoms.* New York: W.H. Freeman, 1998.

Margulis, Lynn, and Dorian Sagan. *The Garden of Microbial Delights.* Dubuque, Iowa: Kendall Hunt Publishing Co., 1993.

Marshall, Elizabeth L. *The Human Genome Project.* New York: Franklin Watts, 1996.

Mauseth, James D. *Plant Anatomy.* Menlo Park, Calf.: Benjamin/Cummings Publishing Co., 1988.

Mearns, Barbara. *Audubon to X'antus.* San Diego, Calf.: Academic Press, 1992.

Moore, David M. *Green Planet: The Story of Plant Life on Earth.* New York: Cambridge University Press, 1982.

Morris, Desmond. *Animal Days.* New York: Morrow, 1979.

Morton, Alan G. *History of the Biological Sciences: An Account of the Development of Botany from Ancient Times to the Present Day.* New York: Academic Press, 1981.

Nebel, Bernard J., and Richard T. Wright. *Environmental Science: The Way the World Works.* Upper Saddle River, N.J.: Prentice Hall, 1998.

Nies, Kevin A. *From Priestess to Physician: Biographies of Women Life Scientists.* Los Angeles, Calf.: California Video Institute, 1996.

Norell, Mark, A., Eugene S. Gaffney, and Lowell Dingus. *Discovering Dinosaurs in the American Museum of Natural History.* New York: Knopf, 1995.

O'Daly, Anne, ed. *Encyclopedia of Life Sciences.* 11 vols. Tarrytown, N.Y.: Marshall Cavendish Corp., 1996.

Postgate, John R. *Microbes and Man.* New York: Cambridge University Press, 2000.

Reader's Digest Editors. *Secrets of the Natural World.* Pleasantville, N.Y.: Reader's Digest Association, 1993.

Reaka-Kudla, Marjorie L., Don E. Wilson, and Edward O. Wilson. *Biodiversity II: Understanding and Protecting Our Biological Resources.* Washington, D.C.: Joseph Henry Press, 1997.

Rensberger, Boyce. *Life Itself.* New York: Oxford University Press, 1996.

Rosenthal, Dorothy Botkin. *Environmental Science Activities.* New York: John Wiley & Sons, 1995.

Ross-McDonald, Malcom, and Robert Prescott-Allen. *Man and Nature: Every Living Thing.* Garden City, N.Y.: Doubleday, 1976.

Sayre, Anne. *Rosalind Franklin and DNA.* New York: W.W. Norton & Co., 1975.

Shearer, Benjamin F., and Barbara Smith Shearer. *Notable Women in the Life Sciences: A Biographical Dictionary.* Westport, Conn.: Greenwood Press, 1996.

Shreeve, Tim. *Discovering Ecology.* New York: American Museum of Natural History, 1982.

Singer, Charles Joseph. *A History of Biology to about the Year 1900.* Ames, Iowa: Iowa State University Press, 1989.

Singleton, Paul. *Bacteria in Biology, Biotechnology and Medicine.* New York: John Wiley & Sons, 1999.

Snedden, Robert. *The History of Genetics.* New York: Thomson Learning, 1995.

Stefoff, Rebecca. *Extinction.* New York: Chelsea House, 1992.

Stephenson, Robert, and Roger Browne. *Exploring Variety of Life.* Austin, Tex.: Raintree Steck-Vaughn, 1993.

Sturtevant, Alfred H. *History of Genetics.* New York: Harper & Row, 1965.

Tesar, Jenny E. *Patterns in Nature: An Overview of the Living World.* Woodbridge, Conn.: Blackbirch Press, 1994.

Tocci, Salvatore. *Biology Projects for Young Scientists.* New York: Franklin Watts, 1999.

Tremain, Ruthven. *The Animal's Who's Who.* New York: Scribner, 1982.

Tyler-Whittle, Michael Sidney. *The Plant Hunters.* New York: Lyons & Burford, 1997.

Verschuuren, Gerard M. *Life Scientists.* North Andover, Mass.: Genesis Publishing Co., 1995.

Wade, Nicholas. *The Science Times Book of Fish.* New York: Lyons Press, 1997.

Wade, Nicholas. *The Science Times Book of Mammals.* New York: Lyons Press, 1999.

Walters, Martin. *Innovations in Biology.* Santa Barbara, Calf.: ABC-CLIO, 1999.

Watson, James D. *The Double Helix: A Personal Account of the Discover of the Structure of DNA.* New York: Scribner, 1998.

Wilson, Edward O. *The Diversity of Life.* Cambridge, Mass.: Belknap Press of Harvard University Press, 1992.

Videocassettes

Attenborough, David. *Life on Earth.* 13 episodes. BBC in association with Warner Brothers & Reiner Moritz Productions. Distributor, Films Inc. Chicago, Ill.: 1978. Videocassette.

Attenborough, David. *The Living Planet.* 12 episodes. BBC/Time-Life Films. Distributor, Ambrose Video Publishing, Inc., N.Y. Videocassette.

Web Sites

ALA (American Library Association):
Science and Technology: Sites for
Children: Biology.
http://www.ala.org/parentspage/greatsites/
science.html#c
(Accessed August 9, 2000).

Anatomy and Science for Kids.
http://kidscience.about.com/kids/kidscience/
msub53.htm
(Accessed August 9, 2000).

ARS (Agricultural Research Service):
Sci4Kids.
http://www.ars.usda.gov/is/kids/
(Accessed August 9, 2000).

Best Science Links for Kids
(Georgia State University).
http://www.gsu.edu/~chevkk/kids.html
(Accessed August 9, 2000).

Cornell University: Cornell Theory Center
Math and Science Gateway.
http://www.tc.cornell.edu/Edu/
MathSciGateway/
(Accessed August 9, 2000).

Defenders of Wildlife: Kids' Planet.
http://www.kidsplanet.org/
(Accessed August 9, 2000).

DLC-ME (Digital Learning Center for
Microbiology Ecology).
http://commtechlab.msu.edu/sites/dlc-me/
index.html
(Accessed August 9, 2000).

The Electronic Zoo.
http://netvet.wustl.edu/e-zoo.htm
(Accessed August 9, 2000).

Explorer: Natural Science.
http://explorer.scrtec.org/explorer/
explorer-db/browse/static/Natural/\Science/
index.html
(Accessed August 9, 2000).

Federal Resources for Educational
Excellence: Science.
http://www.ed.gov/free/
s-scienc.html
(Accessed August 9, 2000).

Fish Biology Just for Kids: Florida
Museum of Natural History.
http://www.flmnh.ufl.edu/fish/Kids/
kids.htm
(Accessed August 9, 2000).

Franklin Institute Online: Science Fairs.
http://www.fi.edu/qanda/spotlight1/
spotlight1.html
(Accessed August 9, 2000).

GO Network: Biology for Kids.
http://www.go.com/WebDir/Family/Kids/
At_school/Science_and_technology/
Biology_for_kids
(Accessed August 9, 2000).

Howard Hughes Medical Institute:
Cool Science for Curious Kids.
http://www.hhmi.org/coolscience/
(Accessed August 9, 2000).

Internet Public Library: Science Fair
Project Resource Guide.
http://www.ipl.org/youth/projectguide/

Internet School Library Media Center:
Life Science for K-12.
http://falcon.jmu.edu/~ramseyil/
lifescience.htm
(Accessed August 9, 2000).

K-12 Education Links for Teachers and
Students (Pollock School).
http://www.ttl.dsu.edu/hansonwa/k12.htm
(Accessed August 9, 2000).

Kapili.com: Biology4Kids! Your Biology
Web Site!.
http://www.kapili.com/biology4kids/
index.html
(Accessed August 9, 2000).

Lawrence Livermore National Laboratory:
Fun Science for Kids.
http://www.llnl.gov/llnl/03education/
science-list.html
(Accessed August 9, 2000).

LearningVista: Kids Vista: Sciences.
http://www.kidsvista.com/Sciences/
index.html
(Accessed August 9, 2000).

Life Science Lesson Plans: Discovery
Channel School.
http://school.discovery.com/lessonplans/
subjects/lifescience.html
(Accessed August 9, 2000).

Life Sciences: Exploratorium's
10 Cool Sites.
http://www.exploratorium.edu/
learning_studio/cool/life.html
(Accessed August 9, 2000).

Lightspan StudyWeb: Science.
http://www.studyweb.com/Science/
(Accessed August 9, 2000).

Lycos Zone Kids' Almanac.
http://infoplease.kids.lycos.com/
science.html
(Accessed August 9, 2000).

Mr. Biology's High School Bio Web Site.
http://www.sc2000).net/~czaremba/
(Accessed August 9, 2000).

Mr. Warner's Cool Science: Life Links.
http://www3.mwis.net/~science/life.htm
(Accessed August 9, 2000).

Naturespace Science Place.
http://www.naturespace.com/
(Accessed August 9, 2000).

NBII (National Biological Information
Infrastructure): Education.
http://www.nbii.gov/education/index.html
(Accessed August 9, 2000).

PBS Kids: Kratts' Creatures.
http://www.pbs.org/kratts/
(Accessed August 9, 2000).

Perry Public Schools: Educational Web
Sites: Science Related Sites.
http://scnc.perry.k12.mi.us/
edlinks.html#Science
(Accessed August 9, 2000).

QUIA! (Quintessential Instructional
Archive) Create Your Own Learning
Activities.
http://www.quia.com/
(Accessed August 9, 2000).

Ranger Rick's Kid's Zone: National
Wildlife Federation.
http://www.nwf.org/nwf/kids/index.html
(Accessed August 9, 2000).

The Science Spot.
http://www.theramp.net/sciencespot/
index.html
(Accessed August 9, 2000).

South Carolina Statewide Systemic
Initiative (SC SSI): Internet Resources:
Math Science Resources.
http://scssi.scetv.org/mims/ssrch2.htm
(Accessed August 9, 2000).

ThinkQuest: BodyQuest.
http://library.thinkquest.org/10348/
(Accessed August 9, 2000).

United States Department of the Interior
Home Page: Kids on the Web.
http://www.doi.gov/kids/
(Accessed August 9, 2000).

USGS (United States Geological Service)
Learning Web: Biological Resources.
http://www.nbs.gov/features/education.html
(Accessed August 9, 2000).

Washington University School of Medicine
Young Scientist Program.
http://medinfo.wustl.edu/~ysp/
(Accessed August 9, 2000).

Index

Italic type *indicates volume number;*
boldface *indicates main entries and
their page numbers;* (ill.) *indicates
photos and illustrations.*

A

Abbe, Ernst *2:* 383

Abdomen *1:* 145

Abiotic/Biotic environment *1:* **1–2,**
180–81

Abscisic acid *3:* 465

Abyssal zone *3:* 425

Acid rain *1:* **4–6**
  effects of, *1:* 6 (ill.)

Acids *1:* 2–4

Acquired immune deficiency syndrome.
  See AIDS (acquired immune deficiency
  syndrome).

Acquired immunity *2:* 316–17

Acrosome *3:* 547

Adaptation *1:* **7–8,** 207, 403

Adaptive radiation *1:* 209

Adenine *1:* 165, 170

Adenosine triphosphate (ATP) *2:* 389

Adrenal glands *1:* 191, 193; *3:* 531, 553

Aerobic/anaerobic *1:* **8–11**

Aerobic respiration *2:* 346; *3:* 514

Agent Orange *3:* 466

Agglutination *1:* 69

Aging *1:* **11–13**

study of *1:* 12–13

Agricultural revolution *1:* 16

Agriculture *1:* **13–17,** 15 (ill.)

**AIDS (acquired immune deficiency syn-
  drome)** *1:* **17–21,** 20 (ill.), *2:* 319

Air pollution *3:* 477

Aldrovandi, Ulisse *1:* 195

Algae *1:* **21–24,** 23 (ill.)

Alimentary canal *1:* 158

Alleles *1:* 168; *3:* 499

Alternation of generations *2:* 354; *3:* 550

Alvarez, Luis Walter *1:* 220

Amber *2:* 239

Amino acids *1:* **24–25,** 24 (ill.); *3:* 492

Ammonification *2:* 411

Amoebas *1:* **25–27,** 26 (ill.); *3:* 496

Amoeboid protozoans *1:* 27

Amphibians *1:* **27–30;** *2:* 288; *3:* 586
  life cycle of, *1:* 29 (ill.)

Anabolic metabolism *2:* 374

Anabolism *2:* 374

Anaerobic. *See* Aerobic/anaerobic.

Anaerobic respiration *2:* 346; *3:* 514

Anaphase *1:* 106

Anatomy *1:* **30–33**

The Anatomy of Plants *1:* 73

Anaximander *1:* 211

Androgens *3:* 532

Aneuploidy *1:* 125

Angiosperms *3:* 472, 525

Animacules *2:* 380